ENERGY POLICIES, POLITICS AND PRICES

RENEWABLE ENERGY INITIATIVES SUPPORTED BY FEDERAL AGENCIES

ENERGY POLICIES, POLITICS AND PRICES

Additional books in this series can be found on Nova's website under the Series tab.

Additional E-books in this series can be found on Nova's website under the E-book tab.

ENERGY SCIENCE, ENGINEERING AND TECHNOLOGY

Additional books in this series can be found on Nova's website under the Series tab.

Additional E-books in this series can be found on Nova's website under the E-book tab.

ENERGY POLICIES, POLITICS AND PRICES

RENEWABLE ENERGY INITIATIVES SUPPORTED BY FEDERAL AGENCIES

RICKEY TAYLOR
AND
ALLEN E. MARTIN
EDITORS

Nova Science Publishers, Inc.
New York

For permission to use material from this book please contact us:
Telephone 631-231-7269; Fax 631-231-8175
Web Site: http://www.novapublishers.com

Library of Congress Cataloging-in-Publication Data

ISBN: 978-1-62081-645-5

Published by Nova Science Publishers, Inc. ✝ New York

Contents

PREFACE

The nation's reliance on imported oil, rising energy costs, and fossil fuels' potential contribution to climate change have renewed the federal focus on renewable energy. Many federal agencies support renewable energy activities, raising congressional concerns about the number and roles of agencies implementing such efforts. This book examines the wide range of federal activities related to renewable energy and the recent increase in these efforts that have raised congressional concerns about the number of agencies implementing such activities, as well as the roles of agencies responsible for regulating and providing funding to various segments of the renewable energy industry.

In: Renewable Energy Initiatives Supported ... ISBN: 978-1-62081-645-5
Editors: R. Taylor and A. E. Martin

Chapter 1

RENEWABLE ENERGY: FEDERAL AGENCIES IMPLEMENT HUNDREDS OF INITIATIVES*

United States Government Accountability Office

WHY GAO DID THIS STUDY

The nation's reliance on imported oil, rising energy costs, and fossil fuels' potential contribution to climate change have renewed the federal focus on renewable energy. Many federal agencies support renewable energy activities, raising congressional concerns about the number and roles of agencies implementing such efforts.

Currently, there is no comprehensive inventory of federal renewable energy initiatives, making it difficult to identify potential fragmentation or duplication.

GAO was asked to (1) identify agencies' renewable energy-related initiatives and (2) examine the federal roles agencies' initiatives supported. GAO collected data from agencies' budget documents, strategic plans, websites, and through agency data requests and interviews with officials from most of the 24 federal agencies subject to the Chief Financial Officers Act of 1990.

---

* This is an edited, reformatted and augmented version of United States Government Accountability Office, Report to the Committee on Homeland Security and Governmental Affairs, U.S. Senate, Publication No. GAO-12-260., dated February 2012.

These 24 agencies accounted for roughly 98 percent of all federal outlays in 2009. GAO developed an inventory of agencies' renewable energy initiatives, including initiatives for which renewable energy was the focus, as well as initiatives for which it was part of a broader effort. GAO focused on developing this inventory, and as such, did not evaluate initiatives based on financial support or other factors.

This report contains no recommendations. In response to the draft report, Commerce agreed with the overall findings as they relate to the Department, while Homeland Security neither agreed nor disagreed with GAO's findings.

The other agencies provided technical or clarifying comments, which GAO incorporated as appropriate, or had no comments.

What GAO Found

Governmentwide, 23 agencies and their 130 subagencies GAO reviewed implemented nearly 700 renewable energy initiatives in fiscal year 2010. The Departments of Defense (DOD), Agriculture (USDA), Energy (DOE), and the Interior were collectively responsible for almost 60 percent of all initiatives.

The initiatives supported a range of renewable energy sources, and the most commonly supported sources were bioenergy, solar, and wind. Also, the initiatives supported a range of public and private sector recipients, but the large majority provided support to the private sector. Many initiatives supported multiple renewable energy sources and types of recipients, while many others targeted support to one source or recipient. Agencies' renewable energy efforts increased in recent years as a result of the provisions of the American Recovery and Reinvestment Act of 2009 and other factors, but the level of future efforts is less certain with the expiration of these provisions and budget constraints.

Across agencies, more than 80 percent of initiatives span four key federal roles— supporting research and development; using renewable energy in vehicle fleets and facilities; providing incentives for commercialization and deployment; and regulation, permitting, and ensuring compliance. Certain agencies led efforts in each federal role: DOE, DOD, and USDA for research and development; DOD, the General Services Administration, and DOE for fleets and facilities; Treasury and USDA for commercialization and deployment; and Interior and the Environmental Protection Agency for regulation, permitting, and compliance.

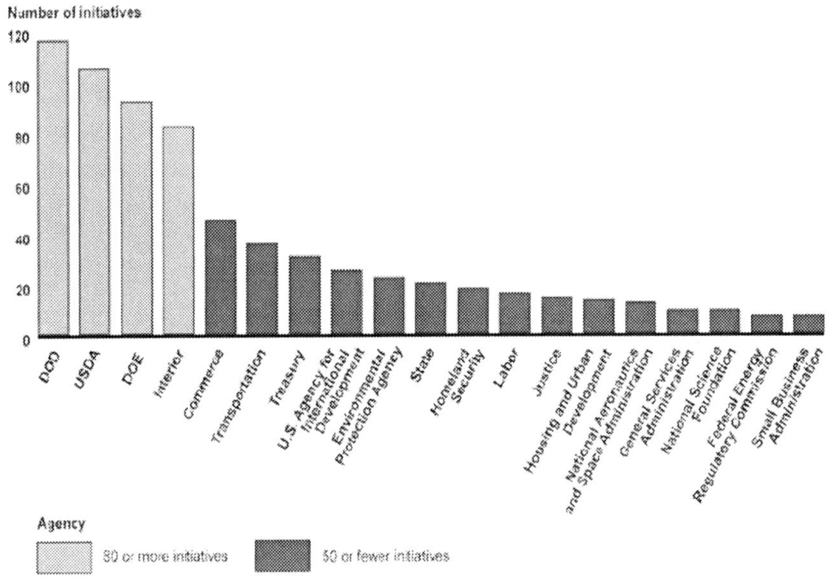

Source: GAI analysis of agencies' data.

Note: Data for DOD include five components—the Air Force, Army, Marine Corps, Navy, and other DOD components that report to the Office of the Secretary of Defense.

Figure. Number of Initiatives by Agency.

ABBREVIATIONS

ARPA-E	Advanced Research Projects Agency-Energy
ARS	Agricultural Research Service
BOEMRE	Bureau of Ocean Energy Management, Regulation and Enforcement
CFO Act	Chief Financial Officers Act of 1990
Commerce	Department of Commerce
DHS	Department of Homeland Security
DOD	Department of Defense
DOE	Department of Energy
DOT	Department of Transportation
EERE	Office of Energy Efficiency and Renewable Energy
EPA	Environmental Protection Agency
EPAct 2005	Energy Policy Act of 2005

FEMP	Federal Energy Management Program
FERC	Federal Energy Regulatory Commission
GREENS	Ground Renewable Expeditionary ENergy System
GSA	General Services Administration
HUD	Department of Housing and Urban Development
Interior	Department of the Interior
IRS	Internal Revenue Service
Justice	Department of Justice
Labor	Department of Labor
NASA	National Aeronautics and Space Administration
NIFA	National Institute of Food and Agriculture
NSF	National Science Foundation
OE	Office of Electricity Delivery and Energy Reliability
Recovery Act	American Recovery and Reinvestment Act of 2009
SBA	Small Business Administration
State	Department of State
Treasury	Department of the Treasury
USAID	U.S. Agency for International Development
USDA	U.S. Department of Agriculture

February 27, 2012
The Honorable Joseph I. Lieberman
Chairman

The Honorable Susan M. Collins
Ranking Member
Committee on Homeland Security and
Governmental Affairs
United States Senate

Americans' daily lives, as well as the economic productivity of the United States, depend on the availability of energy, particularly from fossil fuels. However, concerns over the nation's reliance on imported oil, rising energy costs, and fossil fuels' potential contribution to global climate change have renewed the focus on developing renewable energy resources and technologies to meet future energy needs.[1] According to the Department of Energy's (DOE) Energy Information Administration, domestic renewable energy use grew by 21 percent from 2006 to 2010 and accounted for about 8 percent of U.S.

energy consumption in 2010.[2] During the same period, nuclear energy use increased by about 3 percent, and the use of fossil fuels decreased by about 4 percent largely as a result of the recent economic downturn. Over the coming decades, the Energy Information Administration projects that the strongest growth in the domestic energy sector will be in the use of renewable energy to generate electricity and produce liquid fuels for transportation.

Congress and some federal agencies have emphasized the importance of renewable energy as a means to address issues of national concern and have committed substantial federal resources to initiatives in this area. 3 For instance, DOE's Office of Energy Efficiency and Renewable Energy (EERE) received about $630 million in its fiscal year 2010 appropriation for research related to bioenergy, geothermal, solar, water, and wind energy—an increase of about 43 percent from the fiscal year 2008 level of about $440 million.[4] EERE received an additional $1.4 billion from the American Recovery and Reinvestment Act of 2009 (Recovery Act) for renewable energy-related efforts.[5] The federal government also creates incentives for investment in certain types of renewable energy-related projects and for the production and consumption of renewable energy by, for example, providing favorable tax treatment, such as tax credits, to businesses and individuals. These tax preferences—which are referred to as tax expenditures and are administered by the Department of the Treasury's Internal Revenue Service (IRS)—result in forgone revenue for the federal government.[6] Revenue losses associated with renewable energy-related tax expenditures were estimated to be almost $8.9 billion for fiscal year 2010, according to estimates from the President's 2012 budget.[7] The Recovery Act and other recent legislation established, extended, or changed the scope of a number of these tax expenditures.

The federal government is also uniquely positioned to affect the development of renewable energy resources through its land management and regulatory activities and as a consumer of energy. For example, the Department of the Interior (Interior) manages approximately 500 million acres, or one-fifth, of the nation's land and 1.7 billion acres off its shores. Interior has recently emphasized development of renewable resources in these areas by, for example, implementing measures to streamline the regulatory processes associated with constructing solar energy projects on some of these lands. The federal government is also the single largest user of energy in the United States, and a number of recent federal laws and executive orders have established requirements and direction for federal agencies to (1) reduce energy consumption and greenhouse gas emissions and increase renewable energy use at their facilities, and (2) to reduce petroleum consumption and increase the use of alternative fuels in their vehicle fleets.[8]

The wide range of federal activities related to renewable energy and the recent increase in these efforts have raised congressional concerns about the number of agencies implementing such activities, as well as the roles of agencies responsible for regulating and providing funding to various segments of the renewable energy industry. However, there is currently no comprehensive inventory of which federal agencies are implementing renewable energy-related initiatives and the types of initiatives they are implementing. In light of efforts to balance the federal budget and target spending on activities that will most effectively meet national needs, the lack of available information on agencies' renewable energy initiatives has further raised congressional concerns about the ability to identify whether efforts are fragmented, duplicative, or operating at cross-purposes.

Due to these concerns, you asked us to provide information on federal renewable energy initiatives. Our objectives for this report are to (1) identify federal agencies' renewable energy-related initiatives governmentwide, and (2) examine the federal roles these agencies' initiatives support. Appendix II and GAO-12-259SP provide more detailed information on individual agencies' initiatives.[9] In addition, we provide information on the extent to which the Recovery Act established, expanded, or modified agencies' renewable energy-related initiatives, and identify those initiatives, such as certain tax expenditures that supported ethanol or other biofuels, for which authority has recently or will soon expire without congressional action. This information is provided in appendixes III and IV, respectively.

To identify federal agencies' renewable energy-related initiatives, we collected information on initiatives that were funded, planned, implemented, or authorized in fiscal year 2010 by reviewing agencies' budget documents and other key information sources, such as strategic plans and websites. From this effort, we developed data on agencies' initiatives that were related to renewable energy through a specific emphasis or focus, even if renewable energy was part of a broader effort.[10] We defined an initiative as a program or group of activities serving a similar purpose or function, and the initiatives we identified included agency spending programs as well as tax expenditures. In some instances, these initiatives corresponded to distinct agency programs or initiatives. In other cases, we identified and grouped similar activities into initiatives based on our own judgment when there did not already appear to be a formal name for the initiative, or disaggregated higher-level that went beyond standard governmentwide efforts to incorporate renewable energy into their vehicle fleets and facilities in response to requirements and direction established by federal laws and executive orders.[12] These agencies were: the departments of Agriculture (USDA), Commerce, Defense (DOD), Energy, Homeland Security (DHS), Housing and Urban Development (HUD), the Interior, Justice, Labor, State, Transportation (DOT), and the Treasury; as well as the following independent agencies: the Environmental Protection Agency (EPA), General Services Administration (GSA), National Aeronautics and Space Administration (NASA), National Science Foundation (NSF), Small Business Administration (SBA), and U.S. Agency for International Development (USAID). Of these 18 CFO Act agencies, we collected data on DOD activities separately for the Air Force, Army, Marine Corps, Navy, and other DOD components that report to the Office of the Secretary of Defense. In addition, we collected data for the Federal Energy Regulatory Commission (FERC) separately from DOE.[13] As a result, we provide data on 23 agencies rather than 18.

We submitted a structured data request to agencies to provide information on each of their renewable energy initiatives, including the responsible agency component(s); a description of its purpose, how it is implemented, and how it relates to renewable energy; the applicable renewable energy sources; the recipients of funding, services, or other types of support; the extent to which it was established, expanded, or modified by the Recovery Act; and whether the legislative authority for the initiative has expired or may expire. We then conducted interviews with agency officials responsible for providing the data to collect additional information and assess the accuracy, reliability, and completeness of the data provided. We determined the data to be sufficiently reliable for the purposes of this report. We then analyzed the data we collected to develop descriptive information across the agencies' renewable energy-related initiatives. We recognize that measures other than the number of initiatives implemented, such as the level of financial support provided, are also important to understanding renewable energy efforts governmentwide. An agency may have many initiatives, but the total financial support provided for these initiatives could be substantially less than the financial support provided by an agency with fewer initiatives. We did not review agencies' renewable energy-related initiatives on the basis of the level of financial support agencies provide for their renewable energy efforts because, for example, officials stated that financial support for renewable energy is often not tracked separately from other activities. Therefore, we could not collect reliable renewable energy-specific funding data or revenue loss estimates across the full inventory of initiatives we identified.

To examine the federal roles agencies' renewable energy initiatives support, we analyzed the initiative data we collected to identify the federal roles of the initiatives each agency implemented. Specifically, we categorized agencies' initiatives on the basis of four key federal roles, including (1) research and development; (2) commercialization and deployment; (3) regulation, permitting, and compliance; and (4) fleets and facilities.[14] In some instances, we found that agencies' initiatives did not fit into any of these four roles, and in such instances, we included these initiatives in an "other" category. Appendix I provides a more detailed description of our scope and methodology.

We conducted this performance audit from April 2010 through February 2012 in accordance with generally accepted government auditing standards. Those standards require that we plan and perform the audit to obtain sufficient, appropriate evidence to provide a reasonable basis for our findings and conclusions based on our audit objectives. We believe that the evidence obtained provides a reasonable basis for our findings and conclusions based on our audit objectives.

BACKGROUND

EERE defines renewable energy as "energy derived from resources that are regenerative or for all practical purposes can not be depleted." For purposes of our work, we defined renewable energy sources as follows:

- *Bioenergy.* Energy produced from biological fuel sources in liquid, solid, or gaseous form, not including energy produced from animal, household, or business waste.[15]
- *Geothermal energy.* Energy produced by the internal heat of the earth.
- *Hydroelectric energy.* Energy produced by the force of moving water, excluding ocean environments.
- *Ocean energy.* Energy produced by tides, waves, currents and ocean thermal gradients.
- *Solar energy.* Energy transmitted from the sun.
- *Waste conversion.* Energy generated from waste material from animals, households and businesses, including anaerobic digestion of agricultural or other waste, landfill gas, and municipal solid waste.
- *Wind energy.* Energy produced by the movement of air.

Table 1. Federal Renewable Energy Requirements for Agency Fleets and Facilities

Statute or executive order	Renewable energy requirement
Energy Policy Act of 1992(Oct. 24, 1992)	*Fleet.*Generally requires that agencies acquire alternative fuel vehicles for 75 percent of new light-duty acquisitions by fleets of 20 or more vehicles that are capable of being centrally fueled in metropolitan statistical areas with populations of 250,000 or more.[a]
Energy Policy Act of 2005 (EPAct 2005) (Aug. 8, 2005)	*Facilities.*Generally requires that, to the extent economically feasible and technologically practicable, agencies' renewable electricity consumption meet or exceed 3 percent of totalconsumption in fiscal years 2007 through 2009, 5 percent in fiscal years 2010 through 2012, and 7.5 percent in fiscal year 2013 and thereafter. It also provides additional credit toward these goals for agencies with renewable electricity produced on-site. Finally, it requires the installation of 20,000 solar energy systems in federal buildings by 2010.
Executive Order 13423(Jan. 24, 2007)	*Facilities.*Requires that at least half of all renewable energy required under EPAct 2005 comes from renewable sources placed in service after January 1, 1999. It also requires agencies to the extent feasible to implement renewable energy generation projects on agency property for agency use.
Energy Independence and Security Act of 2007 (Dec. 19, 2007)	Fleet. Generally requires agencies to increase annual alternative fuel consumption by 10 percent by 2015 from a 2005 baseline, and install at least one renewable fuel pump at each federal fuel center by 2010.
Executive Order 13514 (Oct. 5, 2009)	Fleet and facilities. Requires agencies to develop Strategic Sustainability Performance Plans to achieve a fiscal year 2020 percentage reduction target of agencywide greenhouse gas emissions, and consider reductions associated with increasing agency renewable energy use and on-site projects, among other things.

Source: GAO analysis.

[a] Under the Energy Policy Act of 1992, as amended, alternative fuel vehicles include any dedicated, dual-fuel vehicle designed to operate on at least one alternative fuel. In 2008, the act was amended to include hybrid vehicles that use both gasoline and stored energy in a battery to power the vehicle. Alternative fuels under DOE regulations include: methanol, ethanol, and other alcohols; blends of 85 percent or more of alcohol with gasoline; natural gas; liquefied petroleum gas (propane); coal-derived liquid fuels; hydrogen; electricity; biodiesel; and p-series fuels. 10 C.F.R. § 490.2.

Federal agencies implement many renewable energy-related initiatives in response to several key pieces of legislation and federal requirements. For example, USDA received specific direction and authority to implement a number of renewable energy initiatives under the Farm Security and Rural Investment Act of 2002, and the Food, Conservation, and Energy Act of 2008, commonly known as the 2002 and 2008 Farm Bills. Similarly, DOT received authority to implement a number of renewable energy initiatives under the Safe, Accountable, Flexible, Efficient Transportation Equity Act: A Legacy for Users, passed in August 2005. The Recovery Act also initiated, expanded, or modified a broad range of renewable energy activities across the government. In addition, several laws and executive orders established requirements or provided direction to federal agencies to incorporate renewable energy into the operation of their vehicle fleets and facilities, as shown in table 1.

Federal agencies use several methods of financing renewable energy projects on their property to help meet the requirements of these laws and executive orders.[16] In addition to funding the full cost of projects using direct appropriations, agencies may use a number of alternative financing arrangements, whereby financing for the projects comes from parties outside of the federal government, such as private energy services companies. Alternative financing can be provided through various mechanisms, including energy savings performance contracts, through which agencies contract with energy services companies to finance and install renewable energy systems, and agencies repay the companies over time; power purchase agreements, through which a developer installs a renewable energy system on agency property under an agreement that the agency will purchase the power generated by the system; or some other financing agreement whereby an outside party provides funding for the project.

There have been several recent estimates of the level of federal financial support for renewable energy:

- The Energy Information Administration estimated total federal subsidies for renewable sources in fiscal year 2010 to be approximately $14.7 billion, including $8.2 billion in tax expenditures. This amount represented a substantial increase from its estimate for fiscal year 2007 of $5.1 billion going to renewable sources. Much of this increase was due to the Recovery Act, which provided an estimated $6.2 billion of the $14.7 billion in fiscal year 2010 subsidies.[17]
- The Congressional Research Service estimated fiscal year 2010 federal revenue losses and outlays associated with renewable energy-related tax provisions to be approximately $13 billion, including Recovery Act funding of approximately $4.2 billion for Treasury payments for energy projects, primarily renewable energy projects.[18]

- The Environmental Law Institute estimated total federal subsidies for renewable sources over the 7-year period from fiscal years 2002 through 2008 to be approximately $29 billion.[19]

Several factors limit these estimates of federal financial support for renewable energy. For instance, many federal programs or tax expenditures support a wide range of renewable and conventional energy sources, as well as businesses or activities outside of the energy sector. These estimates use varying criteria to determine which programs or tax expenditures to include, and for those they do include, the analyses often do not isolate the financial support for renewable energy provided by programs or tax expenditures that may support a broader range of activities or energy sources. In addition, none of these estimates include the level of financial support provided to the renewable energy industry through federal efforts to incorporate renewable energy in agency vehicle fleets and facilities. These estimates also do not account for the costs to the federal government of implementing regulatory efforts related to renewable energy. Furthermore, these estimates do not include all federal agencies with programs that support renewable energy, and for those they do include, they do not always provide a full list of which programs are included.

Federal and state agencies, as well as public and private utilities, implement initiatives beyond those included in our review. First, at the federal level, agencies outside the scope of our review—agencies not subject to the CFO Act—implement renewable energy-related initiatives. For example, the Overseas Private Investment Corporation partners with the private sector to invest in renewable energy and other markets in foreign countries, and the Export-Import Bank of the United States finances the foreign purchase of domestically-produced goods, including renewable energy technologies. In addition, the Commodity Futures Trading Commission and the Securities and Exchange Commission implement regulatory efforts involving markets for renewable energy certificates and commodities. Second, at the state level, the 50 states and the District of Columbia implement initiatives, including tax incentives, rebates, grants, and loans to individuals, businesses, or local governments for a wide range of renewable energy projects. Third, public and private utilities offer incentives to promote the installation and use of renewable energy systems by their customers, including rebates, loans, and performance-based incentives—payments based on the amount of energy generated by a renewable energy system.

Agencies Implemented Hundreds of Initiatives That Support Renewable Energy

Across the federal government, we identified almost 700 renewable energy-related initiatives for fiscal year 2010 at the agencies and subagencies we examined. We found that initiatives in this inventory supported all of the renewable energy sources in our analysis, with initiatives most commonly supporting bioenergy, solar, and wind. In addition, we found that agencies' initiatives supported a range of recipients in both the public and private sectors, with the majority of initiatives supporting private sector recipients. Given a variety of factors, such as recent legislation and future budget uncertainty, agencies' fiscal year 2010 renewable energy efforts represent an increased level of activity from prior years, but future activity levels are less certain.

Federal Agencies Implemented Almost 700 Initiatives in Fiscal Year 2010

Governmentwide, we identified 679 renewable energy-related initiatives for fiscal year 2010 at the 23 agencies we examined and their 130 subagencies.[20] These agencies fell into two groups based on the number of initiatives they implemented: agencies responsible for more than 80 initiatives each and agencies that had 50 or fewer initiatives. Figure 1 shows the number of initiatives by agency.

Agencies with more than 80 initiatives—DOD, USDA, DOE, and Interior— were collectively responsible for almost 60 percent of all initiatives.

- *DOD.* DOD's components collectively accounted for 116 of the initiatives we identified. At DOD, we found renewable energy initiatives at the Army (30), Air Force (17), Navy (16), and Marine Corps (12). Within each of these services, the renewable energy initiatives we identified were dispersed across a number of civilian and military elements of the chain of command. At the Army, for example, responsibility for energy management activities at its facilities is shared by the Office of the Assistant Secretary of the Army for Installations, Energy and Environment (civilian) and the Office of the Assistant Chief of Staff for Installation Management (military). Other components of the Army share responsibility for a number of research efforts to develop technologies for the Army's use, some of which include renewable energy. Other Army renewable energy initiatives are led by the U.S. Army Corps of Engineers, which is the largest hydropower producer in the United States. In addition to the military services, we identified 41 initiatives across a number of components that report to the Office of the Secretary of Defense, such as the Office of the Under Secretary of Defense for Acquisition, Technology and Logistics' Office of the Assistant Secretary, Operational Energy Plans and Programs; the Defense Advanced Research Projects Agency; and the Defense Logistics Agency.

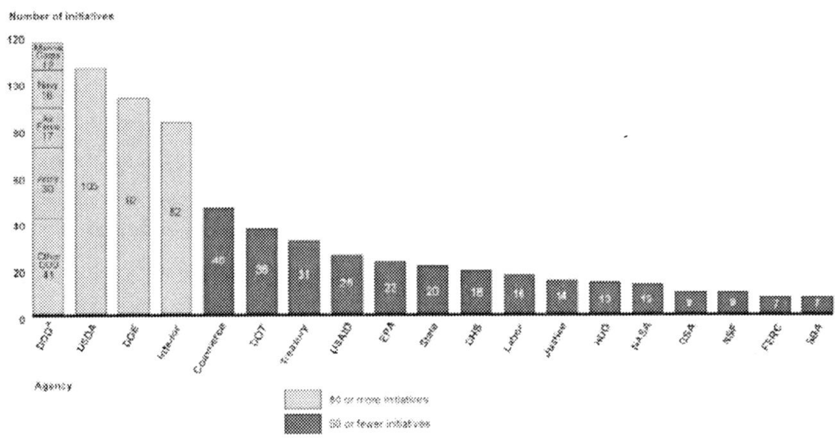

Source: GAO analysis of agencies' data.

Note: The figure includes 5 initiatives for which we did not receive responses from the following agencies to determine whether or not they should be considered renewable energy-related initiatives: the Air Force (1), NASA (3), and the Navy (1). We kept these initiatives for purposes of inclusiveness.

[a]The figure shows data for 19 agencies; however, the DOD data include the Air Force, Army, Marine Corps, Navy, and other DOD components. The total number of initiatives across these DOD components was 116.

Figure 1. Number of Initiatives by Agency.

- *USDA.* USDA accounted for 105 initiatives across 17 subagencies. These subagencies span six of USDA's seven Under Secretaries as well as other elements of USDA. Specifically, we identified renewable energy initiatives at subagencies within the Under Secretaries' mission areas for Research, Education, and Economics

(47); Natural Resources and Environment (24); Farm and Foreign Agricultural Services (9); Rural Development (9); Marketing and Regulatory Programs (6); and Food Safety (1). In addition, we identified 10 initiatives implemented by the Office of the Secretary, the Office of the Chief Economist, or Departmental Management.[21]

- *DOE.* DOE accounted for 92 initiatives across a number of program offices and other subagencies. The program offices included EERE (47); the Office of Science (12); the Office of Electricity Delivery and Energy Reliability (OE) (5); the Advanced Research Projects Agency-Energy (ARPA-E) (4); and the Loan Programs Office (2). Other DOE subagencies at which we identified renewable energy initiatives included the Power Marketing Administrations (8), the Office of Policy and International Affairs (4), and the Energy Information Administration (1).[22] We also identified 9 initiatives for which multiple DOE subagencies were responsible.
- *Interior.* Interior accounted for 82 initiatives, which spanned 8 subagencies across all of Interior's main Assistant Secretaries' offices. These included renewable energy initiatives at subagencies under the Assistant Secretaries for Water and Science (28); Land and Minerals Management (26); Fish, Wildlife and Parks (13); Indian Affairs (12); and Insular Affairs (2). We also found that the Office of the Secretary and the Office of the Assistant Secretary for Policy, Management and Budget, together, had 3 initiatives. [23]

Within the 15 agencies that had 50 or fewer initiatives, we also found that a number of subagencies within each agency were responsible for renewable energy initiatives. We found that the number of these subagencies with renewable energy initiatives ranged from 9 at Commerce and DOT to 3 at GSA and NSF. Table 2 shows the range of the number of subagencies with renewable energy initiatives at agencies with 50 or fewer initiatives.

The number of initiatives that agencies implemented, and the range of their subagencies involved provide some indication of their support for renewable energy overall, but certain agencies with relatively fewer initiatives than others have initiatives that may have a relatively greater impact on the renewable energy sector.[24] For example, Treasury, with 31 initiatives, was not among the 4 agencies with the largest number of initiatives. However, some of its initiatives, such as the renewable energy-related tax expenditures administered by IRS, potentially apply to a wide range of taxpayers eligible to take advantage of the relevant portions of the tax code. As a result, these tax expenditures may provide a greater level of support for renewable energy development than is provided by other types of agency programs, such as grant programs, where the level of support is tied to the level of funding provided through the program.[25] In addition, EPA's Renewable Fuel Standard program, mandated by the Clean Air Act, as amended, generally requires the volume of biofuels used in the transportation sector in the 48 contiguous states to increase through 2022 to an annual total of 36 billion gallons.[26] This requirement applies to all producers and importers of gasoline and diesel and any party that produces, transports, stores, or blends the renewable fuel. As such, its impact is felt by a wide range of businesses across the United States.

Table 2. Number of Subagencies with Renewable Energy Initiatives at Agencies with 50 or Fewer Initiatives

Agency	Number of subagencies
Commerce	9
DOT	9
Treasury	7
Justice	6
Labor	6
DHS	5
FERC	5
HUD	5
NASA	5
USAID	5
EPA	4
SBA	4
State	4
GSA	3
NSF	3

Source: GAO analysis of agencies' data.

Note: As noted previously, the federal agencies within the scope of this report are organized along a range of different structures that include formal subagencies, bureaus, services, divisions, and offices, among other types of structures. For ease of reporting, we generally refer to all of these organizational elements as subagencies.

Initiatives Most Commonly Supported Bioenergy, Solar, and Wind

Among the hundreds of initiatives we identified across federal agencies, we found that initiatives supported all of the renewable energy sources in our analysis, including bioenergy, waste conversion, geothermal, hydropower, ocean, solar, and wind. However, bioenergy, solar, and wind were the most commonly supported energy sources across the inventory. Figure 2 shows the number of initiatives that support each energy source, which sums to more than 679—the number of initiatives in the inventory— because about half of the initiatives supported multiple energy sources.[27]

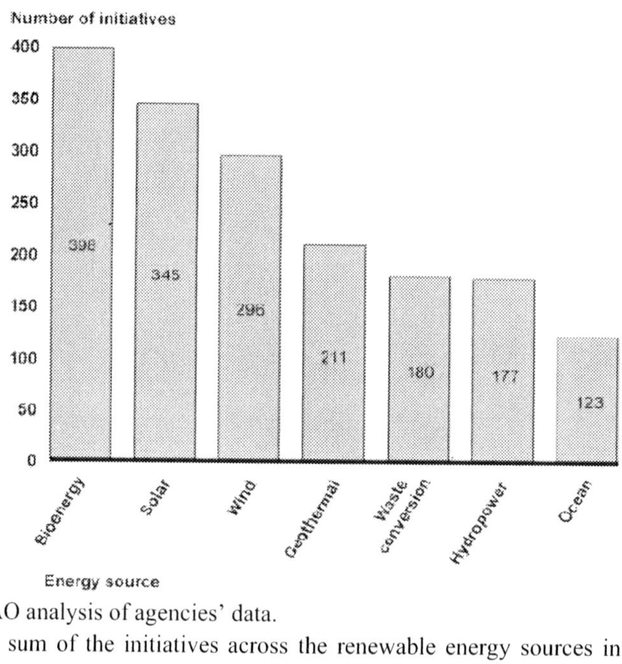

Source: GAO analysis of agencies' data.

Notes: The sum of the initiatives across the renewable energy sources in this figure adds to more than the total number of initiatives in our inventory because initiatives often supported multiple energy sources. Also this figure does not include data for 29 initiatives for which we could not determine the applicable renewable energy sources based on the responses received from the following agencies: the Air Force (1), the Army (1), DOE (15), HUD (4), NASA (3), the Navy (2), the other DOD components that report to the Office of the Secretary of Defense (1), State (1), and USDA (1).

Figure 2. Number of Initiatives Supporting Each Renewable Energy Source.

Across the three renewable energy sources most commonly supported by the renewable energy initiatives identified, we found that (1) certain subcategories of these sources were more commonly supported than others, and (2) certain agencies—USDA, DOD, and Interior—led federal support for these energy sources based on their relative number of initiatives.[28] Specifically:

- *Bioenergy.* Nearly 60 percent of the renewable energy initiatives we identified supported some type of bioenergy as an energy source, including liquid biofuel, solid biomass fuel, or other types of bioenergy. We found that most of the initiatives that supported bioenergy supported liquid biofuel. Of the agencies in our review, USDA had the largest number of initiatives that supported bioenergy—accounting for more than 20 percent of the initiatives that supported bioenergy. USDA's initiatives supported bioenergy in a variety of ways, such as efforts to: (1) develop new types of bioenergy crops, including biofuel or biomass crops or improve processes for producing bioenergy from these crops; (2) expand the planting and harvesting of bioenergy crops and biofuel refining capacity; (3) provide information about biofuel production and market trends; and (4) support the U.S. biofuel industry in the international arena.
- *Solar.* Approximately half of the renewable energy initiatives supported solar energy, including technologies such as photovoltaics, concentrating solar power, and solar hot water.[29] We found that the large majority of the initiatives that supported solar energy supported photovoltaic technologies. Of the agencies in our review, the military services and other components of DOD had the largest number of initiatives that supported solar energy—accounting for more than 20 percent of the initiatives that supported solar energy. DOD's initiatives supported solar energy in various ways and included a number of efforts to incorporate renewable energy at its installations, as well as to develop technologies that allow it to use solar energy in operational environments, such as the conflicts in Iraq and Afghanistan, to help reduce its reliance on conventional fuel.
- *Wind.* More than 40 percent of the renewable energy initiatives supported wind energy, including land-based and off-shore wind energy. We found that the large majority of these initiatives supported land-based wind energy. While there are currently no off-shore commercial scale wind energy projects operating in the United States,

some agencies' initiatives focused on developing the technologies and regulatory framework for off-shore wind energy projects. Of the agencies in our review, Interior had the greatest number of initiatives that supported wind energy—accounting for about 15 percent of the initiatives that supported wind energy. Among other efforts, Interior's wind energy initiatives involved examining the impacts of development, authorizing and permitting wind development on public lands and off-shore areas, and ensuring that projects comply with applicable laws and regulations.

Overall, the initiatives we identified were fairly evenly split between those that supported one energy source and those that supported multiple sources. Initiatives in the inventory targeted certain energy sources or provided support to a broader set of sources in a variety of ways. Table 3 provides examples of those initiatives that supported multiple energy sources and those that were targeted to one source.

Initiatives Most Commonly Supported Private Sector Recipients

We found that agencies' initiatives provided support to a range of recipients in both the public (governmental) and private (generally nongovernmental) sectors; however, the majority of them supported private sector recipients. Specifically, almost 90 percent of the initiatives we identified supported private sector recipients either exclusively or along with public sector recipients. Figure 3 shows the percentage of initiatives that supported private or public sector recipients, or both. Initiatives supported private sector recipients in a wide range of efforts that, for example, funded research or construction of renewable energy projects. Many agencies supported renewable energy research conducted by private companies, universities, or by other nongovernmental researchers. For example, several of the agencies we examined funded renewable energy research as part of their participation in the Small Business Innovation Research program. Through this program, which is administered by SBA, agencies with a budget of $100 million or more for extramural research and development—research that is conducted by nonfederal employees outside federal facilities—are to provide funding to small businesses to, among other efforts, develop commercial products that address their needs.[30] In addition, many agencies we examined implemented renewable power or energy projects at their facilities, such as installing solar arrays, ground-source heat pumps, and wind turbines.[31]

Table 3. Examples of Initiatives That Support a Broad Range of or Are Targeted to Certain Renewable Energy Sources

Name of initiative (agency)	Description
Initiatives that support a broad range of renewable energy sources	
Rural Energy for America Program (USDA)	USDA provides guaranteed loans and grants to farmers, ranchers, and small businesses in rural areas to assist with purchasing and installing renewable energy equipment that could involve all of the energy sources included in our analysis.
Credit for Residential Energy Efficient Property (Treasury)	IRS administers this tax expenditure that provides a credit to homeowners of up to 30 percent of the costs of eligible energy equipment. This credit is focused on several renewable energy sources common to residential properties, such as geothermal, solar, and wind.
Green Jobs Innovation Fund (Labor)	Labor provides competitive grants to help workers receive job training in "green" industry sectors andoccupations, including the renewable energy sector, to ensure that workers have the necessary skills, knowledge, and abilities to succeed in high growth and emerging industries.
Certified Development Company/Section 504 Loans (SBA)	SBA provides loan guarantees to small businesses investing in renewable energy sources or equipment.
Initiatives that are targeted to specific renewable energy sources	
National Institute of Standards and Technology Biomass—Cellulosic Ethanol Research (Commerce)	Institute staff conduct research on potential ways to overcome barriers in the production of cellulosic ethanol.
Aids to Navigation Photovoltaics (DHS)	To make U.S. waterways more navigable, the Coast Guard operates and administers a system of aids to navigation around the nation including buoys, lighthouses, and other lighted structures. Since the late 1950s the Coast Guard has used solar photovoltaic panels to power these structures.
NextGen Environmental Research—Aircraft Technologies, Fuels and Metrics (DOT)	The Federal Aviation Administration conducts research to develop alternative aviation fuels and technologies, including biofuels, and to better understand their air quality and climate impacts as partof its effort to improve the nation's air travel system.
CHE-DMR-DMS Solar Energy Initiative (SOLAR) (NSF)	NSF provides grants to support interdisciplinary research to address the scientific challenges of highly-efficient harvesting, conversion, and storage of solar energy.

Source: GAO summary of agency-provided data.

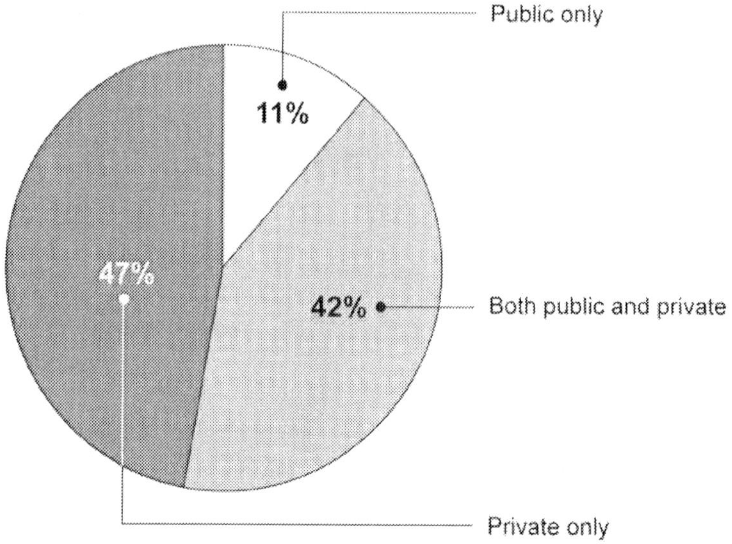

Source: GAO analysis of agencies' data.

Note: This figure does not include data for 11 initiatives for which we could not determine the applicable recipients based on the responses received from the following agencies: the Air Force (1), HUD (4), NASA (3), the Navy (2), and State (1).

Figure 3. Percentage of Initiatives Supporting Private or Public Sector Recipients, or Both.

Generally, these efforts directly supported recipients in the private sector as equipment manufacturers and contractors received the funding to design, construct, operate, and maintain renewable power or energy equipment at agencies' facilities. The smaller percentage of initiatives that supported public sector recipients also supported a range of activities. For example, DOE, DOD, USDA, and other agencies have research laboratories where staff carry out renewable energy research, along with research on other topics. In addition, some agencies' initiatives support state, local, or tribal government activities. For example, EPA provides guidance and oversight for state regulatory programs that include regulating aspects of renewable energy production or use, such as the underground storage of biofuels. In another example, HUD's Neighborhood Stabilization Program awarded grants to state and local governments, among other types of recipients, to assist in the redevelopment of abandoned and foreclosed homes and includes funding for facilitating renewable energy projects, such as photovoltaic panels.

Table 4. Examples of Initiatives That Support a Broad Range of or Are Targeted to Certain Recipients

Name of initiative (agency)	Description
Initiatives that support a broad range of recipients	
Business and Industry Guaranteed Loan Program (USDA)	USDA provides guaranteed loans to borrowers in rural areas in an effort to improve the economic and environmental climate in those areas. Borrowers may include a wide range of public and private entities and individuals.
Climate Technology Initiative – Private Financing Advisory Network (USAID and State)	USAID and State provide technical assistance to project developers and private banks to facilitate financing of renewable energy projects overseas.
Foundational Observation Data Sets (Commerce)	Commerce provides weather observations and forecast data that are generated from its observation networks, including, in some instances, renewable energy resource potential. These observations and data are available to the public, including the renewable energy industry.
Hydropower Licensing Authority (including hydrokinetic resource projects) (FERC)	FERC's responsibilities include the licensing and oversight of hydropower projects. Recipients of this oversight are project applicants and existing licensees, including a broad range of nonfederal governmental and private sector entities.
Initiatives that are targeted to specific types of recipients	
Nano Power Africa (USAID)	USAID partners with universities to create self-sustainable graduate-level programs in science andengineering in sub-Saharan Africa. Participants in these programs conduct research to develop low-cost indigenously produced solar energy devices using nano-materials.
Energy Efficiency Grants for Education and Technology (SBA)	SBA provides grants to small business development centers that, in turn, provide training and counseling services to small businesses. These grants support training in the use and development of renewable energy, among other things.
New Era Rural Technology Program (USDA)	USDA provides grants to community colleges and other post-secondary institutions in rural areas for research and training in the fields of bioenergy and agriculture-based renewable energy resources, among other fields.
People, Prosperity, and the Planet Award Program (EPA)	EPA awards funding to teams of college students who propose research on renewable energy-related and other topics that supports sustainability.

Source: GAO summary of agency-provided data.

Alternatively, some initiatives provided support to tribal governments to enhance their economic self-sufficiency or the skills of their members by, for example, providing funding for renewable energy projects or training on renewable energy technologies and resource management. In addition, while tax expenditures administered by the IRS generally support activities of private sector taxpayers, some tax expenditures also provide support for renewable energy efforts of public sector entities. For example, certain tax expenditures administered by IRS help public sector entities finance renewable energy projects by providing a credit to the bondholders who invest in such projects; thereby allowing the public sector entities to finance their projects with bonds that have lower interest rates than they would otherwise be able to issue.[32]

Similar to initiatives' support of renewable energy sources, we found that, in some instances, agencies' initiatives supported renewable energy activities implemented by a broad range of recipients, while others were more narrowly targeted. Table 4 provides examples of those initiatives that supported multiple recipients and those that were targeted to specific types of recipients.

Agencies' Renewable Energy Efforts Increased Leading up to Fiscal Year 2010, but Future Activity Levels Are Less Certain

Due to a variety of factors, such as recent legislation and future budget uncertainty, the initiatives we identified for fiscal year 2010 represent an overall increased level of agency activity over prior years, and a higher level of activity than may occur in the future. A number of agency officials reported that renewable energy efforts at their agencies have increased in recent years as a result of the Recovery Act. We found that the Recovery Act expanded or modified agencies' renewable energy initiatives in several ways, such as: (1) establishing new initiatives, (2) providing additional funding to existing initiatives, and (3) extending or changing the scope of existing initiatives. For example, the Recovery Act provided additional funding to GSA to procure vehicles for federal agencies—many of which were alternative fuel vehicles—and exchange them for the agencies' less fuel-efficient vehicles.[33] In addition, many officials reported that their agencies' renewable energy efforts had increased in recent years due to other laws or executive orders, some of which provided direction or requirements to increase efforts to incorporate renewable energy into agency vehicle fleets and facilities. A few other officials reported that renewable energy efforts, such as authorizing development on federal lands and off-shore areas, had increased in recent years due to federal legislation. For example, under EPAct 2005, Interior received specific jurisdiction for renewable energy development on the outer continental shelf, which officials reported led to an overall increase in the agency's renewable energy-related efforts. Some officials also reported that their agencies' renewable energy efforts had increased in recent years as a result of the priority placed on renewable energy by their agencies' leadership or by the administration. At the same time, several officials noted uncertainty as to the future level of their agencies' renewable energy efforts due to potential budgetary constraints. Others observed that, although there is interest in increasing renewable energy efforts, such as installing renewable energy projects at their facilities, funding is limited and that they are looking into the possibility of financing these projects through third parties, such as private energy services companies.

Furthermore, authorization for some initiatives, including programs funded by the Recovery Act and certain tax expenditures that supported renewable energy projects during fiscal year 2010 are set to or have already expired.[34] A few officials noted that the expiration of these programs or tax expenditures may reduce future renewable energy efforts.

AGENCIES' INITIATIVES SUPPORT SEVERAL KEY FEDERAL ROLES, AND CERTAIN AGENCIES TEND TO LEAD EFFORTS FOR THESE ROLES

Across the federal government, agencies' renewable energy initiatives supported several key federal roles, and within each of these roles, we found that certain agencies tend to lead federal efforts. Governmentwide, more than 80 percent of agencies' renewable energy initiatives span four key federal roles—conducting and supporting research and development, using energy in agency vehicle fleets and facilities, providing incentives for commercialization and deployment, and issuing regulations and permits and ensuring compliance.[35]

Beyond these key federal roles, the remaining 19 percent of initiatives focused on other efforts, such as the power generation, transmission, and marketing activities of the U.S. Army Corps of Engineers and DOE's Power Marketing Administrations; and the training and capacity building or trade and market development activities of agencies like Commerce, Labor, SBA, State, and USAID.[36] Figure 4 shows the percentage of agencies' renewable energy initiatives according to their federal role and other.

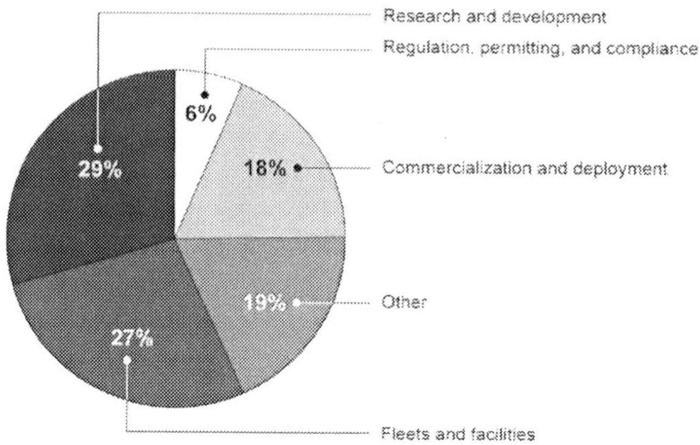

Source: GAO analysis of agencies' data.

Notes: This figure does not include data for 5 initiatives for which we could not determine the applicable federal roles based on the responses received from the following agencies: the Air Force (1), NASA (3), and the Navy (1). Also, percentages do not add to 100 due to rounding.

Figure 4. Percentage of Agencies' Initiatives by Key Federal Role and Other.

At an agency level, we found that each agency's renewable energy initiatives were more frequently concentrated in certain of the four key federal roles we analyzed. Specifically, we analyzed the initiatives to determine which federal roles included a relatively higher percentage of each agency's initiatives. Some agencies' initiatives were concentrated in a single federal role, while other agencies' initiatives were spread across several roles. Table 5 shows the distribution of agencies' initiatives across the four key federal roles.

We identified certain agencies that led efforts within each of the four key federal roles, as follows.

Research and Development

For the research and development key federal role, DOE, DOD, and USDA implemented many initiatives to help develop renewable energy technologies. Specifically:

- *DOE.* Through 51 initiatives, DOE researched and developed what it refers to as "clean energy" technologies—including renewable energy—in support of its mission of ensuring America's security and prosperity by addressing its energy, environmental, and nuclear challenges through transformative science and technology solutions. Several offices within DOE, such as EERE, ARPA-E, OE, and the Office of Science, engage in a wide range of research and development activities that support multiple renewable energy sources. Almost half of DOE's renewable energy research and development initiatives are housed within EERE, which funded 24 initiatives related to, for example, lowering the costs of solar, wind, water, bioenergy, and geothermal technologies, among other renewable energy-related topics. ARPA-E's research and development initiatives invest in efforts to overcome barriers to technology development by providing funding for new high-risk technologies until they are developed enough to obtain private funding or be commercialized.[37] Examples of such technologies related to renewable energy include energy storage options for hydropower and the use of microorganisms to convert carbon dioxide into liquid fuels. In addition, OE's research and development efforts work toward providing a reliable, efficient, and resilient electric power grid, and include developing smart grid and energy storage technologies. EERE, ARPA-E, and OE promote technology-based research for harnessing renewable energy sources, and the Office of Science supports basic science research on diverse topics, including new methods of solar energy capture and conversion and new plant feedstocks—crops or products that can be converted to bioenergy.

- *DOD*. Through 53 initiatives, DOD also engaged in renewable energy-related research and development to support its mission of providing U.S. military forces with the capabilities needed to protect and advance U.S. interests. DOD has determined that renewable energy sources could help achieve its mission by, among other benefits, expanding and securing necessary energy supplies to reduce dependence on foreign oil. Across DOD, each of the military services conducts research efforts that include examining how renewable energy could help meet its particular needs. For example, through their alternative fuel programs, the Air Force and Navy conduct research on new biofuels, biomass, and algae-derived fuels in order to power aircraft and ships. As part of its Alternative Fuels Certification initiative, the Air Force tests and certifies alternative fuels, including renewable alternative fuels such as biomass-derived fuels, and fuel blend technologies in weapons systems (e.g., aircraft), support and fuel delivery equipment, and storage infrastructure. Similarly, the Navy's Alternative Fuels Program focuses on examining whether fuels produced using certain methods can be used in its equipment. These efforts contribute to the high level energy goals set forth in the Navy's Energy Program for Security and Independence, which include sailing the "Great Green Fleet"—a carrier strike group composed of nuclear ships, hybrid electric ships running on biofuel, and aircraft flying on biofuel—by 2016, and meeting at least 50 percent of shore-based energy requirements from alternative sources by 2020, among other energy goals. The Army and Marine Corps also conduct research focused on using renewable energy to meet their deployed forces' needs. For example, the Army's tactical electric power research activities help incorporate renewable energy technologies into tactical electric power generation equipment, such as the Transportable Hybrid Electric Power Station, which uses solar and wind energy to reduce the reliance on fossil fuel generators. In a similar effort, the Marine Corps developed a portable solar power and battery system through the Navy's Future Naval Capabilities Program.

- *USDA*. USDA had 31 renewable energy research and development initiatives to support the nation's food, agriculture, and natural resources supply. For example, USDA's Agricultural Research Service (ARS) conducts a variety of research initiatives that focus on bioenergy feedstocks and related technologies and issues. Under its Bioenergy National Program, ARS develops technologies for the sustainable production of biofuels by the agricultural sector in ways that enhance natural resources without disrupting existing food, feed, and fiber markets. ARS efforts under this initiative include developing new varieties of bioenergy feedstocks, improved production and harvesting practices, and cost-competitive technologies for producing biofuels from feedstocks. As a counterpart to ARS's efforts, the National Institute of Food and Agriculture (NIFA) provides funding for the research efforts of academic researchers and others that focus on developing regional systems of bioenergy production using dedicated energy crops.[38] Its key initiatives include the Agriculture and Food Research Initiative Competitive Grant Program—Sustainable Bioenergy Challenge Area, which provides grants for the sustainable production of bioenergy to help reduce dependence on foreign oil, among other goals. Another key NIFA research and development effort is leading USDA's participation in the Small

Business Innovation Research Program. Across a number of initiatives focusing on biofuels, forests and related resources, rural development, aquaculture, and other areas, NIFA provides grants to small businesses engaged in renewable energy-related research. The Forest Service also conducts research on bioenergy sources, such as forest biomass.

Table 5. Agencies' Initiatives across the Four Key Federal Roles

Agency	Research and development	Fleets and facilities	Commercialization and deployment	Regulation, permitting, and compliance
Commerce	●	●	○	○
DOD				
DOD other	●	●	○	○
Air Force	●	●	○	○
Army	●	●	○	○
Marine Corps	●	●	○	○
Navy	●	●	○	○
DHS	○	●	○	○
DOE	●	○	○	○
DOT	●	○	●	○
EPA	●	○	○	●
FERC	●	●	○	●
GSA	○	●	○	○
HUD	○	○	●	○
Interior	○	○	○	●
Justice	○	●	○	○
Labor	○	○	○	○
NASA	●	●	○	○
NSF	●	○	○	○
SBA	○	○	●	○
State	○	○	○	○
Treasury	○	○	●	○
USAID	○	○	○	○
USDA	●	○	●	○

● Federal role with 25 percent or more of an agency's initiatives

○ Federal role with less than 25 percent of an agency's initiatives

Source: GAO analysis of agencies' data.

Notes: We chose 25 percent as the threshold to examine the distribution of agencies' initiatives across the four key federal roles, assuming that if an agency's initiatives were evenly distributed across the four roles, each of the roles would be darkened. We did not include "other" initiatives in this analysis because we focused on the four key federal roles. Also, this table does not include data for 5 initiatives for which we could not determine the applicable federal roles based on the responses received from the following agencies: the Air Force (1), NASA (3), and the Navy (1).

ARPA-E funds development of cyanobacteria for solar-powered highly efficient production of biofuels

Source: Photos from Arizona State University websites.

One example of the type of research and development efforts supported by DOE's ARPA-E is the $5.2 million grant provided to Arizona State University scientists under the Funding Opportunity Announcement 1 initiative to further develop a renewable source of biodiesel from bacteria known as cyanobacteria. These bacteria use sunlight, water, and carbon dioxide to produce and secrete fat that can be converted to biofuel. According to ARPA-E's website, this novel approach to developing biofuels from microorganisms is more energy efficient than other methods of producing biofuels, which divert more of the sun's energy to cellular growth. In addition, cyanobacteria can be grown on nonarable land, and will not compete for resources with food crops. This 2-year project is a continuation of an earlier $4.7 million Arizona State University pilot project called "Tubes in the Desert," which developed cyanobacteria as a feedstock for biodiesel production, as well as a technology to optimize growth and production of the bacteria using solar energy. This pilot project was completed in 2009 using private funding.

The Marine Corps' Ground Renewable Expeditionary ENergy System

Source: Photo provided by a Marine Corps official.

One example of the Marine Corps' renewable energy research and development activities is the Ground Renewable Expeditionary ENergy System (GREENS), a portable hybrid 300-watt photovoltaic power system. The Marine Corps worked with the Navy to develop this system under the Future Naval Capabilities Program—an effort to leverage research on emerging technologies to close military capability gaps. This effort supports the Marine Corps' focus on meeting operational energy needs with renewable energy. Powered by the sun, GREENS provides continuous electricity for Marines in remote field locations. The GREENS project was conceived in fall 2008 to address military needs in Iraq. The project received expedited approval with technical execution completed in less than 6 months, the first unit tested in July 2009, and solicitation for a production contract in the fall of 2009. In August 2010, GREENS was deployed along with several other renewable energy technologies to Afghanistan for further evaluation under the Experimental Forward Operating Base approach, a Marine Corps initiative to test and evaluate alternative solutions to reduce deployed forces' energy and water needs. According to information provided by the Marine Corps, the tests conducted on renewable energy technologies in Afghanistan yielded promising results.

Fleets and Facilities

For the fleets and facilities key federal role, some agencies such as DOD implemented a large number of initiatives, while others such as GSA and DOE have unique initiatives to guide, track, and coordinate renewable energy efforts. Specifically:

- *DOD.* With 49 initiatives related to nontactical vehicle fleets and facilities and tactical equipment procurement, DOD has the most initiatives in this federal role.

  For example, one initiative implemented by the facilities energy staff within the Deputy Under Secretary of Defense for Installations and Environment is to develop DOD-wide renewable energy policy, provide oversight, and track and report progress for DOD's facilities towards achieving energy goals. Another initiative the facilities energy staff implement is DOD's Energy Conservation Investment Program, a DOD-wide program that provides the military services and DOD agencies with funding for projects selected on the basis of their financial returns, contribution towards energy efficiency and renewable energy goals, and their potential to mitigate energy security risks.

  Apart from these higher level DOD-wide initiatives, each of the military services implements initiatives to carry out its responsibilities for managing individual energy projects at its own installations. Table 6 provides the number of each military service's installations with renewable energy projects.
- *GSA.* GSA plays a key governmentwide support role by constructing, leasing, or managing buildings or other properties on behalf of federal agencies and by serving as the federal government's purchasing agent, which includes integrating renewable energy into federal agencies' vehicle fleets.

  Through its seven fleets and facilities initiatives, GSA has two main responsibilities with regard to renewable energy: (1) to assist with energy procurement for federal facilities and (2) to assist federal agencies in incorporating renewable energy into their vehicle fleets. For example, according to GSA officials, GSA's contracts enable EPA to meet its goal of having all of its electricity come from renewable sources.

GSA officials also stated that the agency helps coordinate efforts across different agencies with energy procurement responsibilities to assist agencies in meeting the energy requirements of federal mandates and executive orders. Similarly, GSA funds renewable energy projects and power purchases and tracks on-site power generation across its own portfolio of buildings.

For vehicle fleets, GSA informs agencies on available vehicle options and capabilities and serves as the mandatory source for all nontactical vehicle purchases by federal agencies. In addition, GSA acts as a primary, but nonmandatory, source for vehicles leased by federal agencies.

Table 6. Number of Each Military Service's Installations with Renewable Energy Projects

	Number of installations with DOD-financed projects			Number of installations with third party-financed projects[a]		
Agency	Installations with small scale projects	Installations with medium scale projects	Installations with large scale projects	Installations with small scale projects	Installations with medium scale projects	Installations with large scale projects
Air Force	>45	7	0	7	6	0
Army	>72	4	0	12	3	4
Marine Corps	13	2	0	6	4	0
Navy	52	6	0	13	5	2

Source: GAO analysis of agencies' data.

Note: Small scale projects generate (or are designed to generate) less than 1,000 megawatt hours of energy per year. Medium scale projects generate (or are designed to generate) from 1,000 to 49,999 megawatt hours per year. Large scale projects generate (or are designed to generate) 50,000 megawatt hours or more per year.

[a]Third party-financed projects are installed using various financing mechanisms, such as energy savings performance contracts, power purchase agreements, or some other financing agreement whereby an outside party, such as a private energy services company, provides funding for the project.

- *DOE.* DOE's Federal Energy Management Program (FEMP) offers technical assistance, guidance, education, and outreach to help federal agencies deploy energy efficiency, renewable energy, and water conservation technologies in an effort to implement cost-effective energy management and investment practices.[39]

With respect to vehicle fleets, FEMP provides guidance and assistance to help agencies implement federal legislative and regulatory requirements that mandate reduced petroleum consumption and increased alternative fuel use. FEMP also helps federal agencies access private sector financing to fund energy efficiency improvements and renewable energy projects at their facilities through energy savings performance contracts, power purchase agreements, and other financing mechanisms. In addition, DOE's Sustainability Performance Office engages in activities designed to implement federal environmental, energy, and transportation management goals at DOE sites.

Commercialization and Deployment

Within the commercialization and deployment key federal role, Treasury and USDA, among other agencies, provided important incentives to stimulate renewable energy development through a variety of different types of financial support. Specifically:

- *Treasury.* Treasury's IRS administered 21 tax expenditures that provided incentives to expand the use of renewable energy in transportation (through increased production and use of biofuels), and electricity or heat generation (through increased investment and manufacturing), among other incentives.

  For example, IRS provides an excise tax credit of 45 cents per gallon to gasoline suppliers and marketers who blend ethanol with petroleum-based fuels.[40] IRS also supports the development of renewable energy for electricity through the Energy Production Tax credit, which, in calendar year 2010, was largely used to develop wind energy, according to information provided by IRS.

  The amount of the credit varies depending upon the source: in calendar year 2010, the credit amount for wind, geothermal, and certain biomass electricity production was 2.2 cents per kilowatt hour, and 1.1 cents per kilowatt hour for electricity produced from other renewable energy sources.

  Table 7 shows the 21 renewable energy-related tax expenditures administered by IRS.

Table 7. Tax Expenditures That Support Renewable Energy

Type of incentive	Tax expenditure name
Transportation incentives	Ethanol Excise and Income Tax Credits [a,c]
	Alcohol and Alcohol Fuel Mixtures Excise and Income Tax Credits [b,c]
	Alternative Fuel Excise Tax Credit[c]
	Small Ethanol Producer Credit[c]
	Credit for the Production of Cellulosic Biofuel
	Biodiesel Excise and Income Tax Credits[c]
	Small Agri-Biodiesel Producer Credit[c]
	Credit for Clean Fuel Burning Vehicles[c]
	Credit for Clean-Fuel Burning Vehicle Refueling Property c
Electricity or heat incentives	Energy Production Credit
	Exclusion of Utility Conservation Subsidies
	Credit for Residential Energy Efficient Property
	Qualifying Advanced Energy Project Credit
	Energy Investment Credit
	Credit for Holding Clean Renewable Energy Bonds
	Credit for Holding New Clean Renewable Energy Bonds
	Direct Payment in Lieu of a Credit for Holding New Clean Renewable Energy Bonds
	Credit for Holding Qualified Energy Conservation Bonds
	Direct Payment in Lieu of a Credit for Holding Qualified EnergyConservation Bonds
	Exclusion of Interest on Energy Facility Bonds (private activity bonds)
	Accelerated Depreciation Recovery Periods for Specific Energy Property

Source: GAO summary of agency data.

Note: This list of tax expenditures was developed based on information provided by IRS officials to represent provisions of the Internal Revenue Code that provide support for renewable energy. The list was developed specifically to represent the criteria for collecting data on distinct renewable energy initiatives for this report. As a result, the count does not necessarily track Treasury's published tax expenditure list which, for example, may encompass multiple provisions in a tax expenditure listed.

[a]The ethanol excise tax credit is known as the Volumetric Ethanol Excise Tax Credit.

[b]The alcohol and alcohol fuel mixtures excise and income tax credits apply to alcohol fuels other than ethanol.

[c]These tax expenditures, or, in some instances, certain components of them were set to expire on December 31, 2011, according to information provided by IRS. As detailed in appendix IV, agencies provided information on expiring initiatives as of fall 2011 and we did not update this information to reflect whether or not any of the initiatives expected to expire prior to the issuance of our final report, such as certain tax expenditures that supported ethanol or other biofuels, actually did expire.

Payments for Specific Energy Property in Lieu of Tax Credits

Treasury's commercialization and deployment initiatives are primarily tax expenditures administered by the IRS, but Treasury was also charged with implementing a cash grant program under Section 1603 of the Recovery Act. Through this program, Treasury provides payments of up to 30 percent of a project's total eligible costs to applicants who place specified energy property, including renewable energy systems, in service for use in a trade or business. Treasury implements this program in coordination with DOE's National Renewable Energy Laboratory, which assists in reviewing the applications. Program applicants may take the payment in lieu of either the Energy Production or Investment Credits administered by IRS, and these payments provide an incentive for investment in property for electricity and heat production, particularly for those applicants without sufficient tax liability to utilize one of the tax credits. Initially, the program provided cash grants for renewable energy projects placed in service as of December 31, 2010, or which began construction in 2009 or 2010. However, the program was extended for 1 year as part of the Tax Relief, Unemployment Insurance Reauthorization, and Job Creation Act of 2010. Under current law, projects have to be under construction by the end of calendar year 2011, with applications submitted by October 1, 2012, to take advantage of the program. For those projects that were awarded funding through October 2011, 97 percent of the projects supported were solar energy projects, while wind energy projects received 79 percent of the funding provided because, according to a Treasury official, these projects tend to be more expensive.

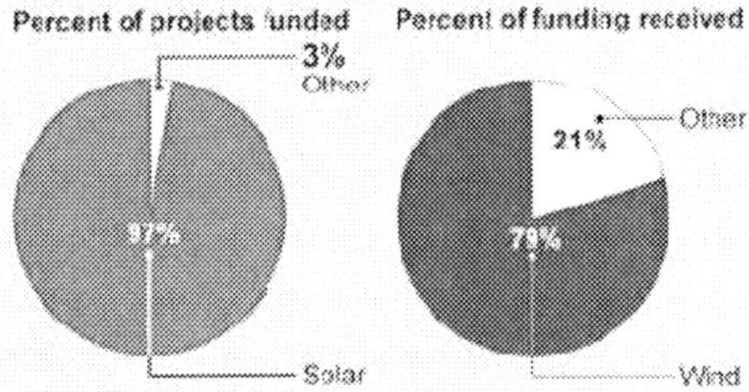

Source: GAO analysis of Treasury data.

Note: In the figures, 'other' includes renewable energy sources such as bioenergy and geothermal, as well as nonrenewable energy sources such as fuel cells and combined heat and power.

In addition to tax expenditures, under Section 1603 of the Recovery Act, Treasury offers payments for specific energy property in lieu of the Energy Production and Energy Investment tax credits. These payments provide an incentive for investment in property for electricity and heat production, particularly those applicants without sufficient tax liability to utilize a tax credit.

- *USDA.* USDA leads the greatest number of commercialization and deployment efforts among federal agencies. USDA had 28 commercialization and deployment initiatives, most of which supported the production of ethanol or other types of bioenergy. Within USDA, the Farm Service Agency, Forest Service, Rural Business Service, and Rural Utilities Service provided incentives for a wide range of activities from producing biofuel feedstocks and building biorefineries to installing renewable energy generation projects on farms. For example, the Farm Service Agency's initiatives focused on supporting the production of biomass crops for energy. Its key initiative—the Biomass Crop Assistance Program—provides payments to farmers and foresters for biomass feedstock production. In addition, the Farm Service Agency is working with other USDA agencies to develop a roadmap to meet Renewable Fuel Standard goals for biofuels production, while the Rural Business Service provided loan guarantees to support the development, construction, or retrofitting of commercial scale biorefineries. The Forest Service also supported commercialization and deployment of bioenergy through a variety of initiatives that provided incentives to utilize biomass collected from Forest Service and state and private lands for energy. Other USDA initiatives supported a broader range of renewable energy sources. For example, the Rural Business Service's Rural Energy for America Program provides guaranteed loans and grants to assist farmers, ranchers, and small businesses to install renewable energy systems. Most fiscal year 2010 renewable energy projects were solar or small-scale wind projects, but many other types of projects, such as biomass anaerobic digestors, were also funded. In addition, the Rural Utilities Service provided loans, loan guarantees, or grants to a variety of recipients to support projects that incorporate renewable energy to improve electric service in rural areas and to assist rural communities with high energy costs.

Regulation, Permitting, and Compliance

Accounting for 6 percent of initiatives overall, initiatives within the regulation, permitting, and compliance key federal role—implemented by several key agencies such as Interior and EPA—affect a wide range of renewable energy development activities.[41] Specifically:

- *Interior.* Interior had 24 initiatives related to regulation, permitting, and compliance of renewable energy projects on federal property—over half of the total number of initiatives across agencies related to this federal role. Interior's Bureau of Land Management regulates solar, wind, and geothermal energy development on public lands. For example, the Bureau of Land Management conducted large-scale solar resource assessments, environmental impact studies, and took other steps to streamline its regulatory processes to expedite solar development on its lands. In addition, in fiscal year 2010, Interior's Bureau of Ocean Energy Management, Regulation and Enforcement (BOEMRE) was responsible for implementing a program to grant leases, easements, and rights-of-way for orderly, safe, and environmentally responsible renewable energy development, such as siting and constructing off-shore wind facilities on the outer continental shelf—generally beyond 3 miles from shore in most states.[42] BOEMRE's initiatives included holding intergovernmental task force meetings on siting for potential off-shore wind projects and conducting stakeholder outreach and coordination, and examining related technologies to understand their environmental impacts and develop standards and inspection guidelines. Recently, BOEMRE increased efforts—particularly for off-shore wind—in part due to heightened interest at the state level, according to officials.
- *EPA.* EPA also has a key role in implementing renewable energy-related regulation, permitting, and compliance activities to meet the objectives of federal legislation to protect the nation's air, water, and human health. For example, under its Renewable Fuel Standard program, EPA's Office of Air and Radiation establishes requirements and ensures compliance on the volume and types of biofuels used in the transportation sector, as required by the Energy Independence and Security Act of 2007. The Office of Air and Radiation also conducts vehicle emissions testing to support future regulation of biofuels emissions. In addition, the Office of Air and Radiation is responsible

for oversight of state and local permitting authorities that issue permits to large industrial sources of pollution, such as factories and power plants, and ensure that pollution from these sources does not exceed limits imposed under the Clean Air Act. As part of this effort, EPA encourages these sources to consider using renewable energy to reduce emissions. EPA's Office of Underground Storage Tanks also performs a regulatory function by overseeing and coordinating with state regulators and private fuel companies regarding the underground storage of biofuels.

For more detailed information on each agency's renewable energy initiatives, see GAO-12-259SP. This e-supplement provides information on the components of each agency responsible for implementing renewable energy initiatives; a description of each initiative, including what it is, how it works, and how it relates to renewable energy; as well as data on the renewable energy sources, recipient types, and federal roles supported by each initiative. Readers who would like analyses of agencies' renewable energy initiatives other than those presented in this report may download the data and sort it in different ways.

AGENCY COMMENTS AND OUR EVALUATION

We provided Commerce, DHS, DOD, DOE, DOT, EPA, FERC, GSA, HUD, Interior, Justice, Labor, NASA, NSF, SBA, State, Treasury, USDA, and USAID with a draft of this report for review and comment. Commerce and DHS provided written comments in which Commerce generally agreed with the overall findings of the report as they relate to the Department of Commerce, and DHS neither agreed nor disagreed with our findings, but provided information on the Department's energy efficiency efforts. In addition, Commerce, DOE, DOT, EPA, HUD, Interior, Labor, NASA, NSF, SBA, and USDA provided technical or clarifying comments, which we incorporated as appropriate. In particular, as part of its technical comments, NSF had one comment of a more general nature which was that the report's title could imply excessive or wasteful federal efforts. In this regard, the report title conveys our primary finding, which is that federal agencies implement hundreds of renewable energy-related initiatives. We did not analyze the extent to which any of these initiatives might be considered excessive or wasteful. We made no change for this comment. DOD, FERC, GSA, Justice, State, Treasury, and USAID indicated they had no comments on the report.

Frank Rusco
Director,
Natural Resources and Environment

APPENDIX I: SCOPE AND METHODOLOGY

This appendix provides information on the scope of our work and the methodology used to (1) identify federal agencies' renewable energy-related initiatives governmentwide and (2) examine the federal roles these agencies' initiatives support. It also provides information on the steps taken to examine the extent to which the American Recovery and Reinvestment Act of 2009 (Recovery Act) established, expanded, or modified agencies' renewable energy-related initiatives, and to identify those initiatives for which authority has recently or will soon expire without congressional action.

To identify federal agencies' renewable energy-related initiatives, we performed a series of steps to determine which agencies implement renewable energy efforts, and what should be reported as a renewable energy-related initiative. We then compiled an inventory of federal agencies' renewable energy-related initiatives by collecting data (1) from agency documents and other information sources, (2) through a structured data request provided to the agencies, and (3) through interviews with agency officials.

At the outset of our work, we determined that it would not be feasible to examine every federal department, agency, or other entity to collect data on renewable energy efforts because many agencies indicated that data on such efforts were not tracked in a centralized, comprehensive, or consistent manner.[1] Therefore, we began by focusing on the 24 agencies subject to the Chief Financial Officers Act of 1990 (CFO Act), as these agencies were collectively responsible for 98 percent of federal outlays in fiscal year 2009.[2] Among these 24 agencies, we further focused our work on those with renewable energy-related activities that went beyond standard governmentwide efforts to incorporate renewable energy into their vehicle fleets and facilities in response to requirements and direction established by federal laws and executive orders. Specifically, we identified a number of the CFO Act agencies which, based on their missions, seemed unlikely to implement renewable energy-related efforts beyond vehicle fleets and facilities efforts. We reviewed these agencies' fiscal year 2011 congressional budget justification documents and strategic plans by performing keyword searches for renewable energy-related terms. We examined the text where renewable energy-related information was identified, followed up with agency officials and, as result, excluded the Department of Health and Human Services, Nuclear Regulatory Commission, Office of Personnel Management, and Social Security Administration from our work. In addition, while we originally included the departments of Education and Veterans Affairs in our scope, we later learned that they did not implement renewable energy efforts beyond vehicle fleets and facilities efforts and, therefore, we also excluded them. Through these efforts we found that 18 of the 24 CFO agencies had activities beyond incorporating renewable energy into their fleets and facilities. Within the Department of Defense (DOD), we collected data separately for each of the military services and from other DOD components that report to the Office of the Secretary of Defense. We also collected data for the Federal Energy Regulatory Commission separately from the Department of Energy (DOE).[3] Therefore, our report provides data on the following 23 agencies:

- Department of Agriculture,
- Department of the Air Force,
- Department of the Army,
- Department of Commerce,
- Department of Defense (other components that report to the Office of the Secretary of Defense),
- Department of Energy,
- Department of Homeland Security,[4]
- Department of Housing and Urban Development,
- Department of the Interior,
- Department of Justice,
- Department of Labor,
- Department of the Navy (Navy),
- Department of the Navy (U.S. Marine Corps),
- Department of State,[5]
- Department of Transportation,
- Department of the Treasury,
- Environmental Protection Agency,
- Federal Energy Regulatory Commission,
- General Services Administration,
- National Aeronautics and Space Administration,

- National Science Foundation,
- Small Business Administration, and
- U.S. Agency for International Development.[6]

To develop a definition of renewable energy we reviewed definitions used in legislation, by federal agencies, and by other organizations. We found that there were a number of similarities as well as some differences among the definitions. For example, while the definitions we reviewed generally defined renewable energy as regenerative and included energy sources such as solar or wind, some definitions considered hydrogen to be a renewable energy source while others did not. Also, definitions did not consistently include or classify waste, such as municipal solid waste, as a renewable energy source. Based on our review of these definitions, we defined renewable energy as energy derived from any of the following sources: bioenergy, such as liquid biofuel and solid biomass fuel; geothermal; hydropower; solar; wind; ocean energy, including wave, tidal, current, and ocean thermal energy; and waste conversion sources, such as landfill gas, and municipal solid waste. This energy can be applied in any form, including electricity, transportation, heating, or small scale applications such as streetlights or trash compactors.[7]

We defined a renewable energy-related initiative as a program or group of agency activities serving a similar purpose or function that involved renewable energy through a specific emphasis or focus, even if renewable energy was only one part of a broader effort.[8] These initiatives included both agency spending programs and tax expenditures. In some instances, initiatives we identified corresponded to distinct agency programs or initiatives. In other cases, we identified and grouped similar activities into initiatives based on our own judgment when there did not already appear to be a formal name for the initiative, or disaggregated higher-level activities that included multiple initiatives. Where we identified individual renewable energy-related projects, such as a specific facility energy effort or grant award, we did not consider the projects to be initiatives themselves; rather we identified the broader program or area of effort to which they belonged as an initiative. As a result, the data that we developed do not necessarily represent initiatives that are of similar size in terms of agency financial commitments, number of projects involved, or other quantitative measures.[9] Also, as a result of this methodology, initiatives that supported or regulated renewable energy, but that were not specifically focused on renewable energy, were excluded from our inventory. For instance, the Department of the Treasury's Internal Revenue Service administers a number of tax expenditures that provide incentives for certain activities, such as research or agricultural activities, that could involve renewable energy but which do not have a specific focus on renewable energy. Other federal agencies have broad regulatory roles that involve renewable energy, but do not have renewable energy-focused initiatives. For instance, within the Department of Agriculture, the Animal and Plant Health Inspection Service's Biotechnology Regulatory Services regulates genetically engineered plants that companies wish to commercialize. These genetically engineered plants may include biofuel crops, but the program has no particular focus on biofuel crops. Similarly, the Department of Transportation's Pipeline and Hazardous Materials Safety Administration has specific jurisdiction over the transportation of biofuels by pipeline. However, it exercises oversight of biofuel pipelines through its existing regulatory and enforcement program addressing all pipeline integrity and safety issues, rather than developing a program that focuses specifically on biofuel transportation.

Using these definitions of renewable energy and a renewable energy-related initiative, and focusing on the 23 agencies included in our review, we developed data on agencies' initiatives that were funded, planned, implemented, or authorized in fiscal year 2010. We did so by using a similar search approach as in our scoping work to examine agencies' fiscal year 2010 and 2011 budget documents, strategic planning documents, other information sources, and agency websites.[10] For each agency, we developed a structured data request that included information on that agency's renewable energy-related initiatives as identified during our keyword searches. We sent the data request to officials from each agency for them to confirm, delete, or add initiatives as appropriate to ensure that the list of initiatives represented those agency activities that incorporated renewable energy in some specific way, and that were funded, planned, implemented, or authorized in fiscal year 2010. As part of the data request, we also collected the following information on each initiative, among other data requested:

- the component(s) of the agency responsible for implementing the initiatives;
- descriptions of the initiatives, including their purpose, how they are implemented, and how they relate to renewable energy;
- the renewable energy sources applicable to the initiatives;
- the direct recipients of funding, services, or other types of support from the initiatives;[11]
- the extent to which the Recovery Act established, expanded, or modified initiatives; and

- issues related to expiring legislative authority to implement the initiatives.

After receiving responses to the data request, we conducted interviews with agency officials responsible for providing the data to (1) assess the accuracy, reliability, and completeness of the data provided, and (2) collect additional data concerning the agencies' renewable energy-related initiatives, such as qualitative information on trends in the agencies' initiatives over time and the availability of data on the level of financial support specifically provided for renewable energy through the initiatives. We determined the data to be sufficiently reliable for the purposes of this report. We then analyzed the data we collected to develop descriptive information on the agencies' renewable energy-related initiatives. Appendix II and GAO-12-259SP provide more detailed information on individual agencies' initiatives. We also summarized the information collected on the extent to which the Recovery Act established, expanded, or modified initiatives and issues related to expiring legislative authority to implement initiatives. Information on these topics is provided in appendixes III and IV, respectively.

The number of initiatives agencies implement and related descriptive information are important to understanding renewable energy efforts governmentwide; however, we recognize that other measures of agencies' activities are important as well. For example, data on the level of financial support agencies provide, coordination of efforts, and other information are also indicators of agencies' renewable energy efforts. During this review, we did not evaluate agencies' renewable energy-related initiatives on the basis of these other indicators. In particular, we did not collect information on the level of financial support agencies provide for their renewable energy efforts because, during follow-up discussions, officials stated that for many initiatives, financial support for renewable energy efforts is not tracked separately from other activities; therefore, we could not collect reliable funding data across the full inventory of initiatives we identified. We recognize that initiatives may vary greatly in the scale of their funding or the number of entities expected to benefit, among other things; however, in this report, we do not discuss these differences or compare the scale of the initiatives we identified.

To examine the federal roles agencies' renewable energy initiatives support, we analyzed the data we collected to identify the federal roles of the initiatives each agency implemented. Specifically, we categorized agencies' initiatives on the basis of whether each initiative was most closely related to four key federal roles or "other" that we identified and defined for purposes of this work as follows:

- *Research and development.* Efforts that further the knowledge of or the ability to create, develop, or improve renewable energy technologies through, for example, grants or other forms of financial support to carry out such work. We included technology demonstration as part of this area for purposes of our analysis.
- *Commercialization and deployment.* Activities that provide incentives for the implementation or promote the competitiveness of renewable energy technologies in the commercial market place. The activities include, for example, loans, grants, and other financial support for the design and construction of renewable energy projects or nonfinancial project-specific assistance, such as technical assistance or legal support.
- *Regulation, permitting, and compliance.* Activities that ensure compliance with federal laws and regulations regarding renewable energy production, use, technologies, or facilities. These activities include, for example, processing applications, conducting compliance reviews and environmental assessments, inspections, and enforcement.
- *Fleets and facilities.* Efforts to incorporate renewable energy into an agency's property, buildings, vehicles, and other transportation equipment. These efforts include, for example, installing renewable energy generation equipment at agency facilities or leasing alternative fuel vehicles. In addition, we included equipment, such as tactical military equipment, that is procured and deployed by agencies and that incorporates renewable energy technologies as part of the fleets and facilities federal role.
- *Other.* Other types of agency efforts related to renewable energy that, for example, provide information, services, or other types of support.

We conducted this performance audit from April 2010 through February 2012 in accordance with generally accepted government auditing standards. Those standards require that we plan and perform the audit to obtain sufficient, appropriate evidence to provide a reasonable basis for our findings and conclusions based on our audit objectives. We believe that the evidence obtained provides a reasonable basis for our findings and conclusions based on our audit objectives.

APPENDIX II: INFORMATION ON FEDERAL AGENCIES' RENEWABLE ENERGY-RELATED INITIATIVES

This appendix provides supplementary information to the data represented in figures and tables in the report. Figures 5 through 96 present a series of four figures per agency. The four figures for each agency show: (1) the number of agency initiatives supporting each renewable energy source; (2) the number of energy sources supported by the agency's initiatives; (3) the percentage of agency initiatives that support public sector or private sector recipients, or both; and (4) the percentage of initiatives for each agency that support each key federal role considered in our analysis or that we categorized as "other."[1] In addition, tables 8 through 14 show the number of agencies' initiatives across the four key federal roles, according to energy source. Each table includes information for initiatives that target a single energy source as well as those that support multiple sources.

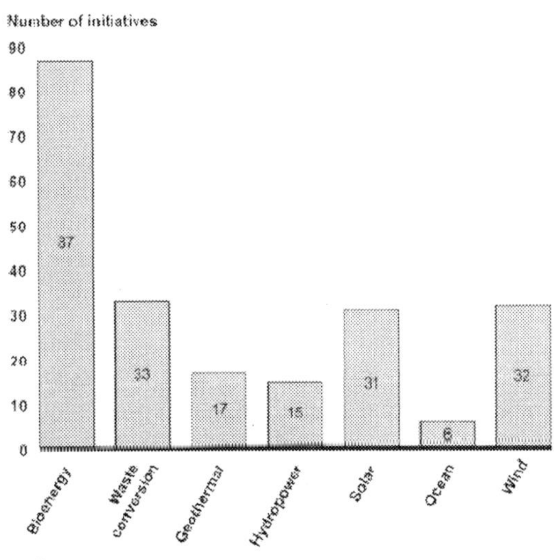

Source: GAO analysis of agency data.

Notes: The sum of the initiatives across the renewable energy sources adds to more than the total number of USDA initiatives because initiatives often supported multiple energy sources. Also, this figure does not include one initiative for which we could not determine the applicable renewable energy sources based on the data USDA provided.

Figure 5. Number of USDA Initiatives Supporting Each Renewable Energy Source.

U.S. Department of Agriculture

The U.S. Department of Agriculture's (USDA) 105 initiatives supported every renewable energy source in our analysis, with a large majority of its initiatives supporting bioenergy, as figure 5 demonstrates. Moreover, figure 6 shows that about half of USDA's initiatives targeted bioenergy exclusively, with a smaller number of initiatives targeted to other energy sources or multiple energy sources. Alternatively, figure 7 shows that USDA's initiatives generally supported either private sector recipients alone or recipients in both the public and private sectors, and figure 8 indicates that USDA's initiatives were distributed across several federal roles, with a much smaller percentage of initiatives in the regulation, permitting, and compliance federal role.

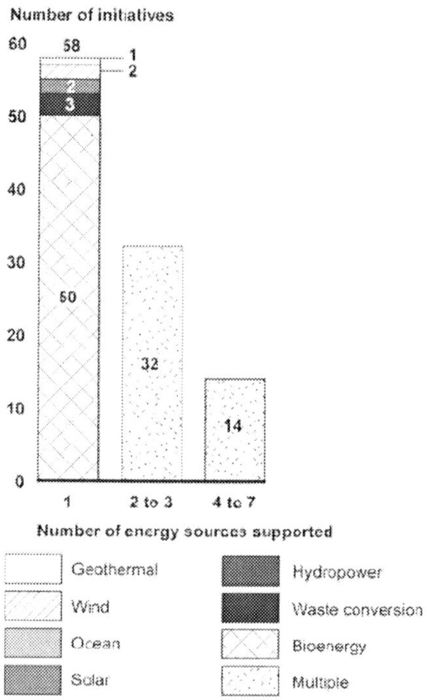

Source: GAO analysis of agency data.

Note: This figure does not include one initiative for which we could not determine the applicable renewable energy sources based on the data USDA provided.

Figure 6. Number of Energy Sources Supported by USDA Initiatives.

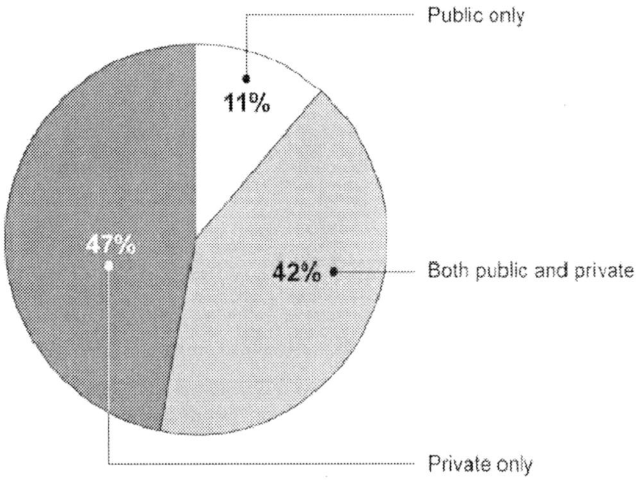

Source: GAO analysis of agency data.

Notes: This figure provides data on the recipients for the 105 USDA renewable energy initiatives we identified. We generally classified governmental recipients as public and nongovernmental recipients as private.

Figure 7. Percentage of USDA Initiatives Supporting Private or Public Sector Recipients, or Both.

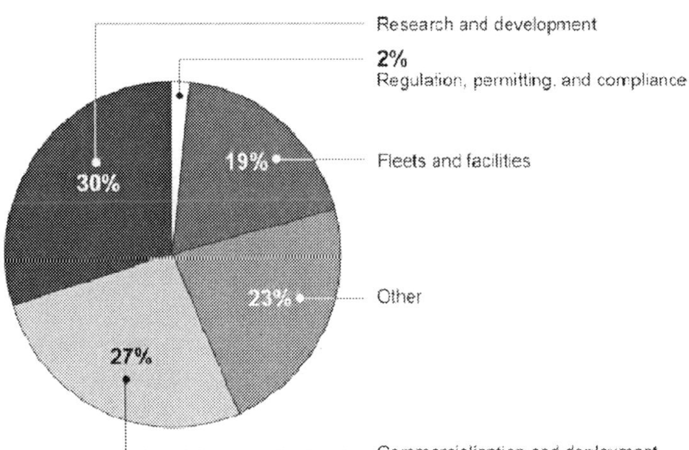

Source: GAO analysis of agency data.

Notes: This figure provides data on the federal roles for the 105 USDA renewable energy initiatives we identified. Percentages do not add to 100 due to rounding.

Figure 8. Percentage of USDA Initiatives by Federal Role and Other.

Department of Commerce

The Department of Commerce's (Commerce) 45 initiatives supported all of the energy sources in our review, with wind, solar, and bioenergy supported the most, as figure 9 shows. In addition, figure 10 shows that roughly half of Commerce's initiatives targeted specific energy sources, with bioenergy the most commonly targeted energy source. Commerce initiatives generally supported either private sector recipients alone or recipients in both the public and private sectors, as demonstrated in figure 11, while figure 12 shows the initiatives supported all the key federal roles in our review, with the highest percentage supporting fleets and facilities.

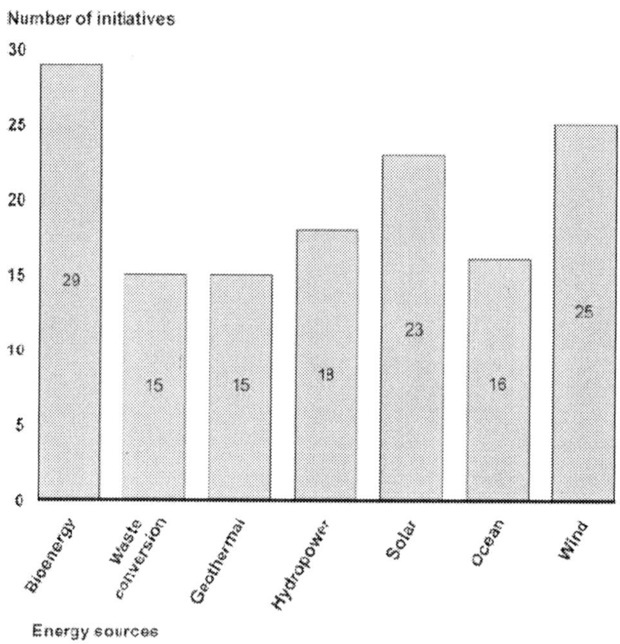

Source: GAO analysis of agency data.

Note: The sum of the initiatives across the renewable energy sources adds to more than the total number of Commerce initiatives because initiatives often supported multiple energy sources.

Figure 9. Number of Commerce Initiatives Supporting Each Renewable Energy Source.

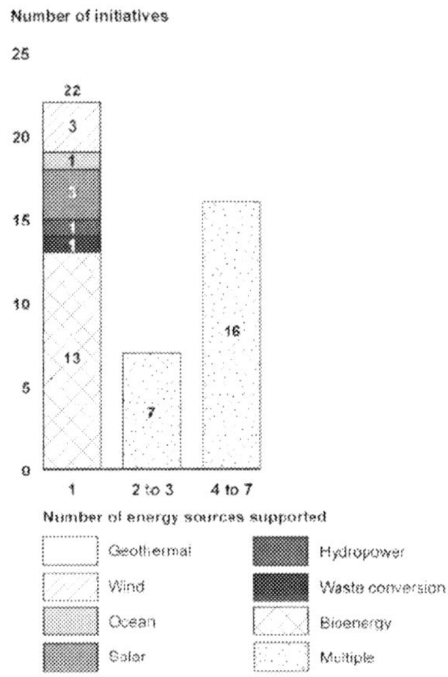

Source: GAO analysis of agency data.

Figure 10. Number of Energy Sources Supported by Commerce Initiatives.

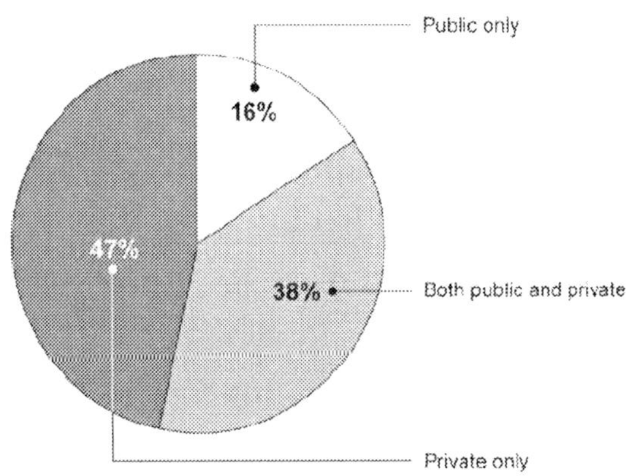

Source: GAO analysis of agency data.

Notes: This figure provides data on the recipients for the 45 Commerce renewable energy initiatives we identified. We generally classified governmental recipients as public and nongovernmental recipients as private. Percentages do not add to 100 due to rounding.

Figure 11. Percentage of Commerce Initiatives Supporting Private or Public Sector Recipients, or Both.

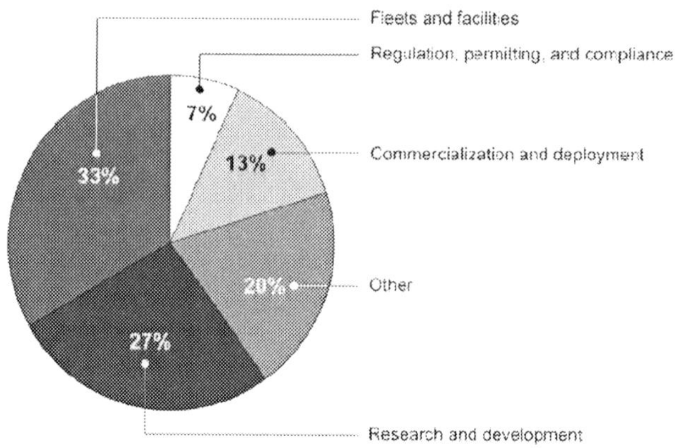

Source: GAO analysis of agency data.

Note: This figure provides data on the federal roles for the 45 Commerce renewable energy initiatives we identified.

Figure 12. Percentage of Commerce Initiatives by Federal Role and Other.

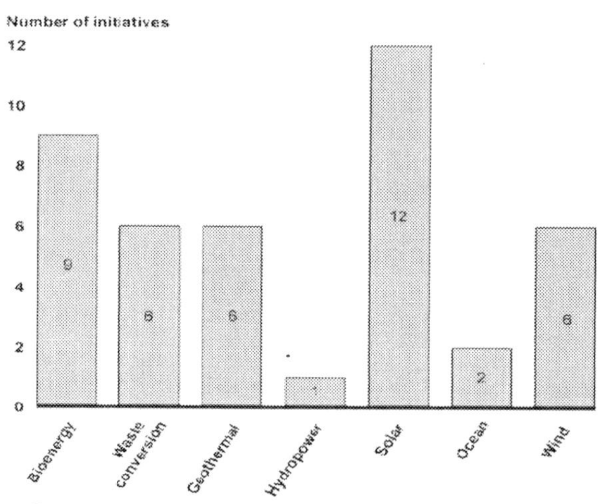

Source: GAO analysis of agency data.

Notes: The sum of the initiatives across the renewable energy sources adds to more than the total number of Air Force initiatives because initiatives often supported multiple energy sources. Also, this figure does not include one initiative for which we could not determine the applicable renewable energy sources based on the data the Air Force provided.

Figure 13. Number of Air Force Initiatives Supporting Each Renewable Energy Source.

Department of Defense, Air Force

The Air Force's 17 initiatives supported all of the energy sources in our review, with solar energy supported more than any other source, as figure 13 shows. Figure 14 demonstrates that the majority of the Air Force initiatives supported more than one energy source, while figure 15 shows that most initiatives supported recipients in both the public and private sectors. As shown in figure 16, half of the Air Force initiatives supported the research and development federal role and most of the other Air Force initiatives supported fleets and facilities.

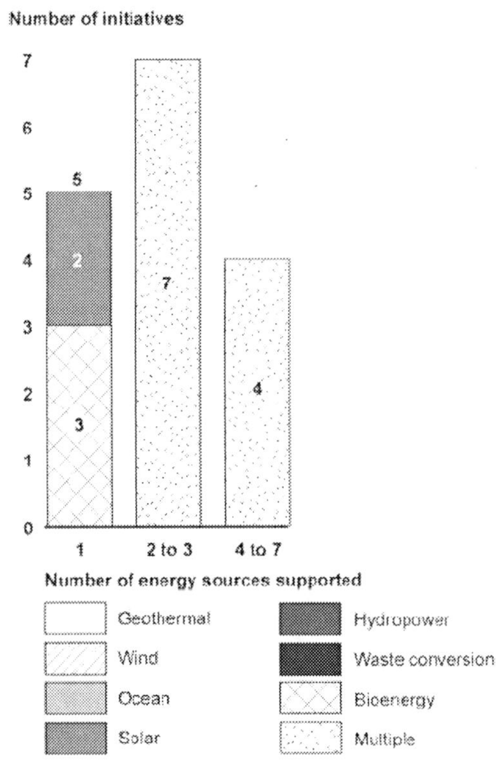

Source: GAO analysis of agency data.

Note: This figure does not include one initiative for which we could not determine the applicable renewable energy sources based on the data the Air Force provided.

Figure 14. Number of Energy Sources Supported by Air Force Initiatives.

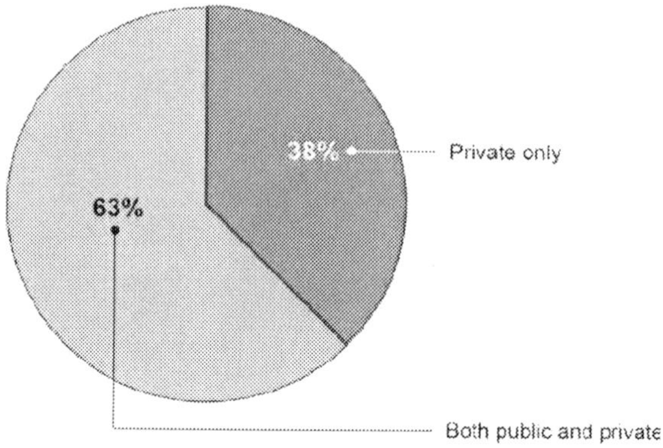

Source: GAO analysis of agency data.

Notes: This figure provides data on the recipients for 16 of the 17 Air Force renewable energy initiatives we identified. The figure does not include one initiative for which we could not determine the applicable recipients based on the data the Air Force provided. We generally classified governmental recipients as public and nongovernmental recipients as private. Percentages do not add to 100 due to rounding.

Figure 15. Percentage of Air Force Initiatives Supporting Private or Public Sector Recipients, or Both.

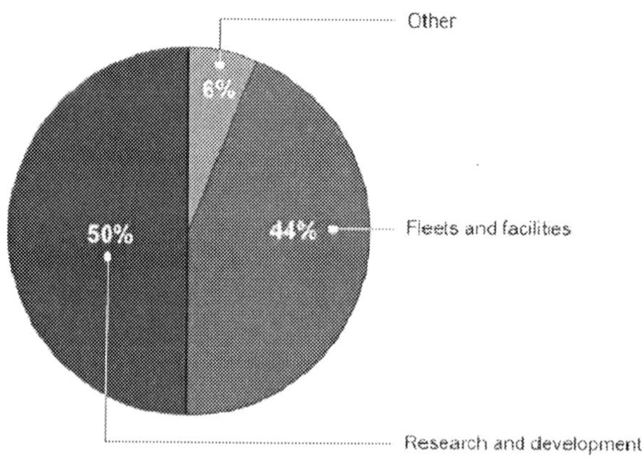

Source: GAO analysis of agency data.

Notes: This figure provides data on the federal roles for 16 of the 17 Air Force renewable energy initiatives we identified. The figure does not include one initiative for which we could not determine the applicable federal role based on the data the Air Force provided.

Figure 16. Percentage of Air Force Initiatives by Federal Role and Other.

Department of Defense, Army

The Army's 30 initiatives supported every renewable energy source in our analysis, with solar, bioenergy, and waste conversion receiving the most support, as figure 17 shows. In addition, as figure 18 shows, more than half of the Army's initiatives supported multiple energy sources, with solar energy receiving the most support from targeted initiatives. As figure 19 demonstrates, 60 percent of the Army's initiatives supported private sector recipients alone. Moreover, the vast majority of the Army's initiatives supported two federal roles, shown in figure 20, with more than half supporting research and development and almost one-third supporting fleets and facilities efforts.

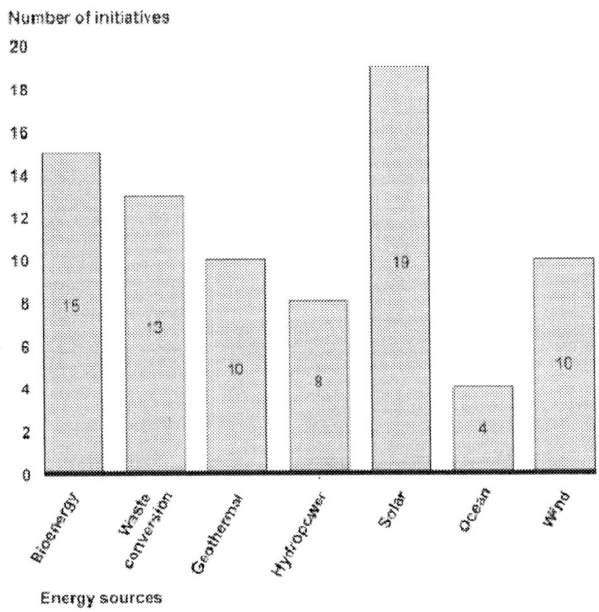

Source: GAO analysis of agency data.

Notes: The sum of the initiatives across the renewable energy sources adds to more than the total number of Army initiatives because initiatives often supported multiple energy sources. Also, this figure does not include one initiative for which we could not determine the applicable renewable energy sources based on the data the Army provided.

Figure 17. Number of Army Initiatives Supporting Each Renewable Energy Source.

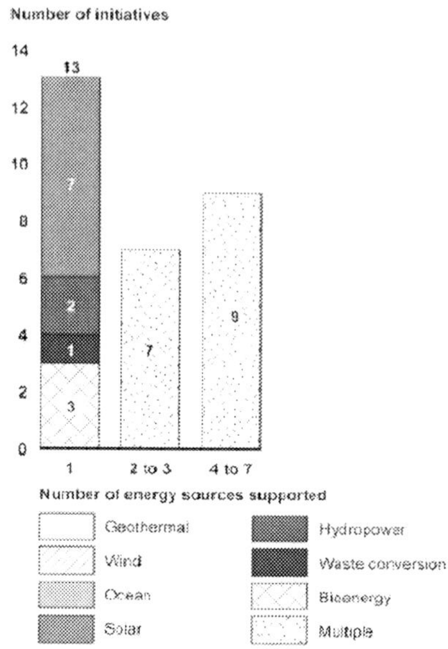

Source: GAO analysis of agency data.

Note: This figure does not include one initiative for which we could not determine the applicable renewable energy sources based on the data the Army provided.

Figure 18. Number of Energy Sources Supported by Army Initiatives.

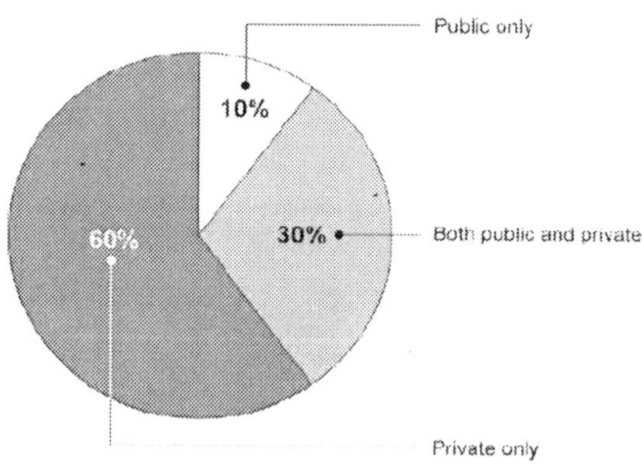

Source: GAO analysis of agency data.

Notes: This figure provides data on the recipients for the 30 Army renewable energy initiatives we identified. We generally classified governmental recipients as public and nongovernmental recipients as private.

Figure 19. Percentage of Army Initiatives Supporting Private or Public Sector Recipients, or Both.

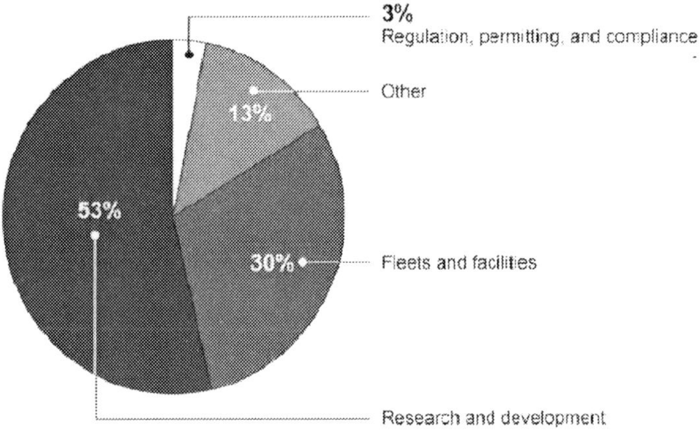

Source: GAO analysis of agency data.

Notes: This figure provides data on the federal roles for the 30 Army renewable energy initiatives we identified. Percentages do not add to 100 due to rounding.

Figure 20. Percentage of Army Initiatives by Federal Role and Other.

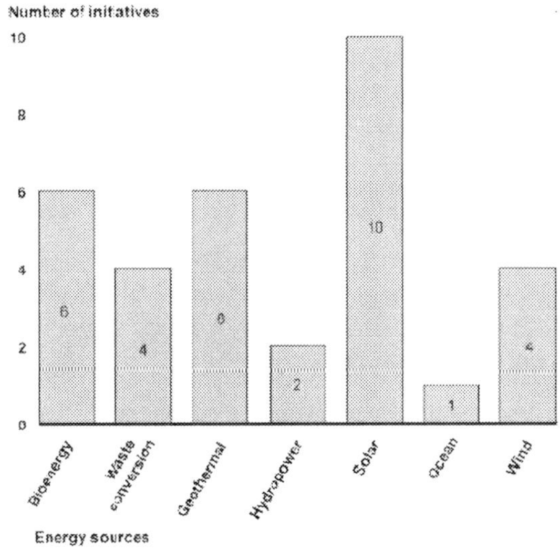

Source: GAO analysis of agency data.

Note: The sum of the initiatives across the renewable energy sources adds to more than the total number of Marine Corps initiatives because initiatives often supported multiple energy sources.

Figure 21. Number of Marine Corps Initiatives Supporting Each Renewable Energy Source.

Department of Defense, Marine Corps

The Marine Corps' 12 initiatives supported all of the energy sources in our review, with solar power supported by the largest number of initiatives, as figure 21 demonstrates. Figure 22 shows that more than half of the Marine Corps initiatives did not target a specific energy source, with those that did target a single source focusing on solar and bioenergy. Almost 60 percent of the initiatives supported recipients in the private sector only, as seen in figure 23, and almost all of the initiatives supported the federal roles of fleets and facilities as well as research and development, as shown in figure 24.

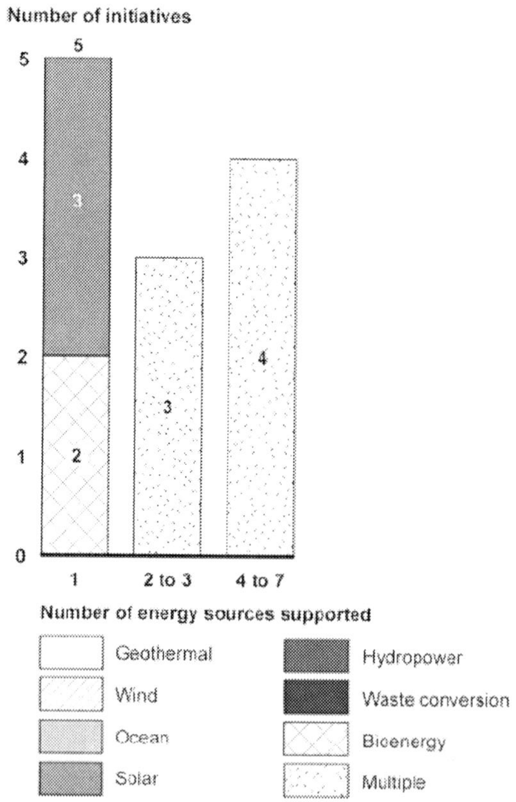

Source: GAO analysis of agency data.

Figure 22. Number of Energy Sources Supported by Marine Corps Initiatives.

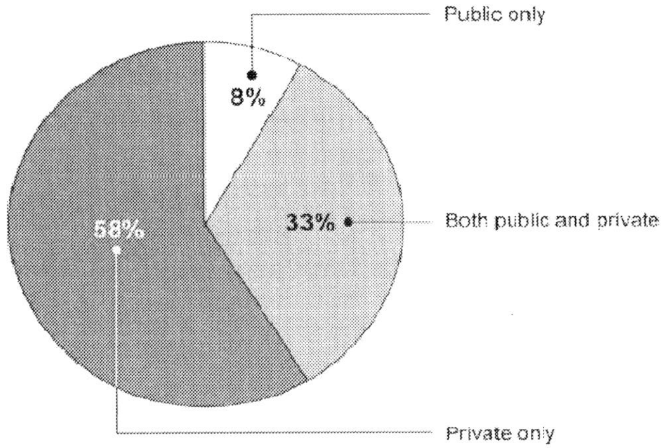

Source: GAO analysis of agency data.

Notes: This figure provides data on the recipients for the 12 Marine Corps renewable energy initiatives we identified. We generally classified governmental recipients as public and nongovernmental recipients as private. Percentages do not add to 100 due to rounding.

Figure 23. Percentage of Marine Corps Initiatives Supporting Private or Public Sector Recipients, or Both.

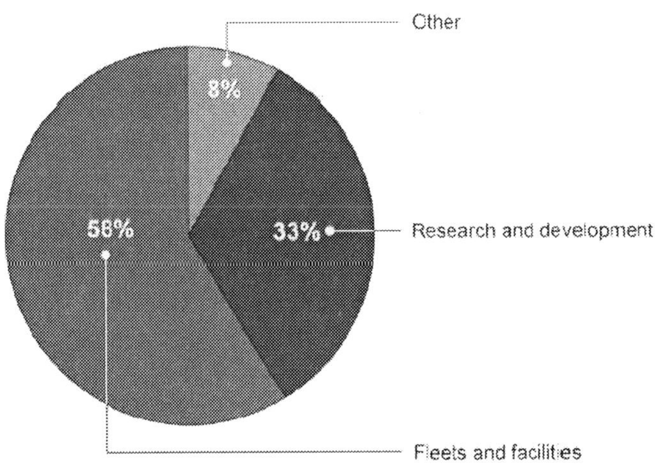

Source: GAO analysis of agency data.

Notes: This figure provides data on the federal roles for the 12 Marine Corps renewable energy initiatives we identified. Percentages do not add to 100 due to rounding.

Figure 24. Percentage of Marine Corps Initiatives by Federal Role and Other.

Department of Defense, Navy

The Navy's 16 initiatives primarily supported solar, bioenergy, and geothermal power, as figure 25 shows. Although more than half the Navy initiatives supported more than one energy source, most of its targeted initiatives focused on bioenergy, as figure 26 indicates. In addition, figure 27 shows that half of the Navy initiatives supported recipients in the private sector alone, while the other half supported recipients in both the public and private sectors. The Navy initiatives were distributed primarily across two federal roles, shown in figure 28, with fleets and facilities supported by more than 50 percent of initiatives and research and development supported by most of the rest.

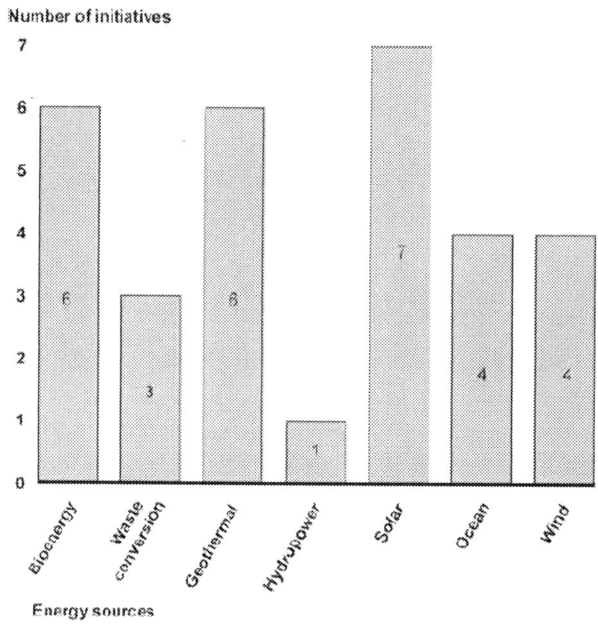

Source: GAO analysis of agency data.

Notes: The sum of the initiatives across the renewable energy sources adds to more than the total number of Navy initiatives because initiatives often supported multiple energy sources. Also, this figure does not include two initiatives for which we could not determine the applicable renewable energy sources based on the data the Navy provided.

Figure 25. Number of Navy Initiatives Supporting Each Renewable Energy Source.

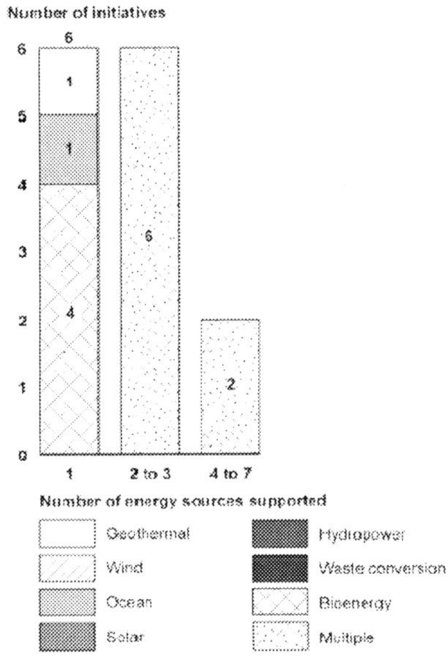

Source: GAO analysis of agency data.

Note: This figure does not include two initiatives for which we could not determine the applicable renewable energy sources based on the data the Navy provided.

Figure 26. Number of Energy Sources Supported by Navy Initiatives.

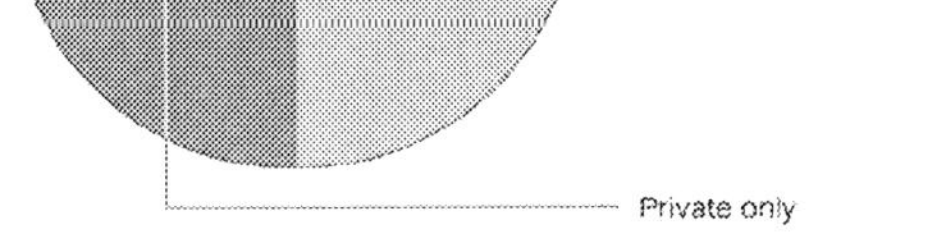

Source: GAO analysis of agency data.

Notes: This figure provides data on the recipients for 14 of the 16 Navy renewable energy initiatives we identified. The figure does not include two initiatives for which we could not determine the applicable recipients based on the data the Navy provided. We generally classified governmental recipients as public and nongovernmental recipients as private.

Figure 27. Percentage of Navy Initiatives Supporting Private or Public Sector Recipients, or Both.

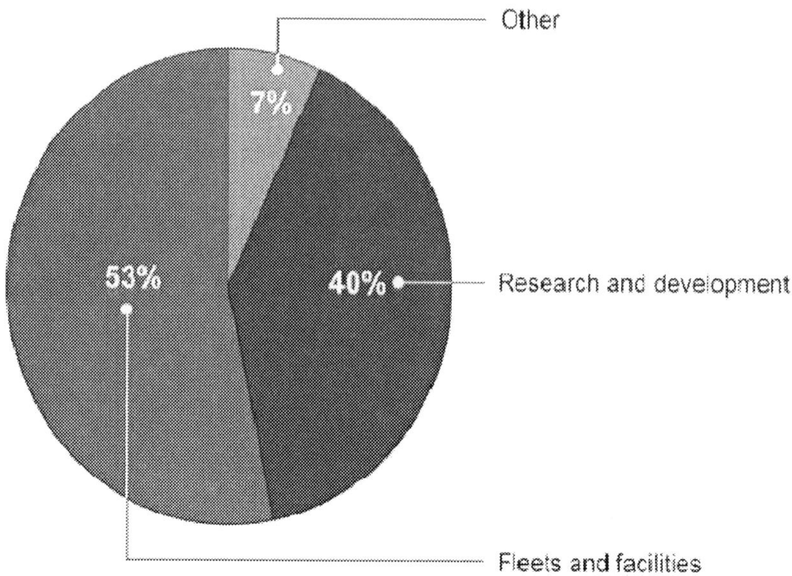

Source: GAO analysis of agency data.

Notes: This figure provides data on the federal roles for 15 of the 16 Navy renewable energy initiatives we identified. The figure does not include one initiative for which we could not determine the applicable federal role based on the data the Navy provided.

Figure 28. Percentage of Navy Initiatives by Federal Role and Other.

Department of Defense, Other Department of Defense Components

The other Department of Defense (DOD) components' 41 initiatives supported all of the energy sources in our review, as figure 29 demonstrates, although solar and bioenergy received the most support. These two energy sources were also targeted the most by initiatives that supported a single energy source, as figure 30 shows.

In addition, figure 31 shows that about two-thirds of the other DOD components' initiatives were targeted to the private sector alone. The other DOD component initiatives were fairly evenly split between the two federal roles of research and development and fleets and facilities, as figure 32 demonstrates.

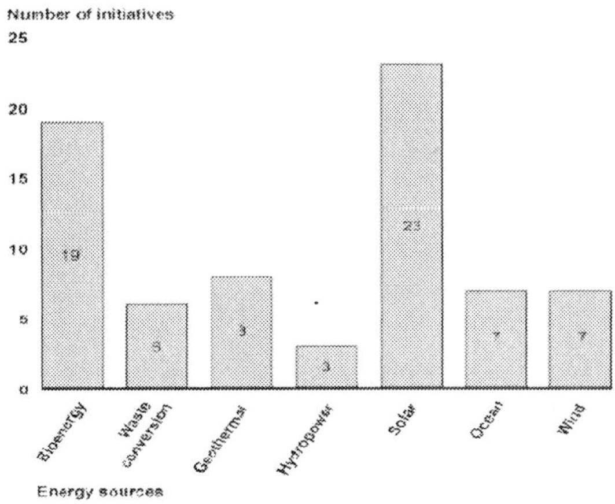

Source: GAO analysis of agency data.

Notes: The sum of the initiatives across the renewable energy sources adds to more than the total number of other DOD component initiatives because some initiatives supported multiple energy sources. Also, this figure does not include one initiative for which we could not determine the applicable renewable energy sources based on the data DOD provided.

Figure 29. Number of Other DOD Component Initiatives Supporting Each Renewable Energy Source.

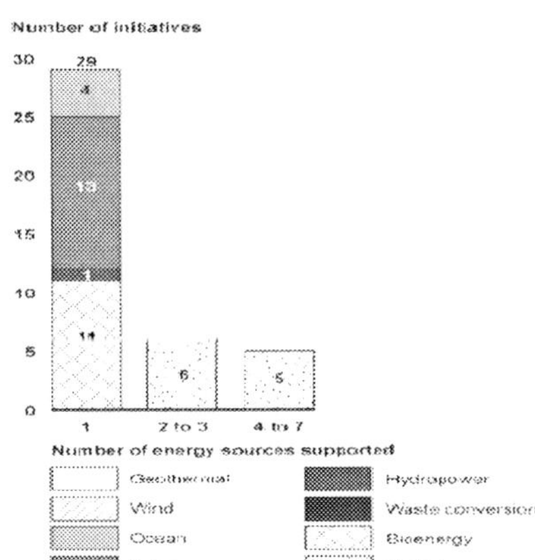

Source: GAO analysis of agency data.

Note: This figure does not include one initiative for which we could not determine the applicable renewable energy sources based on the data DOD provided.

Figure 30. Number of Energy Sources Supported by Other DOD Component Initiatives.

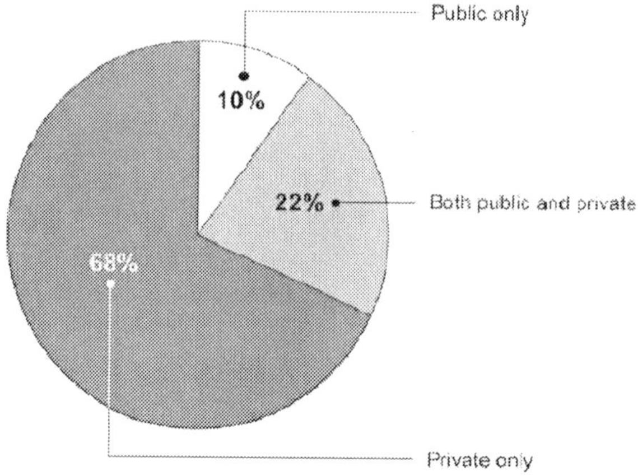

Source: GAO analysis of agency data.

Notes: This figure provides data on the recipients for the 41 other DOD component renewable energy initiatives we identified. We generally classified governmental recipients as public and nongovernmental recipients as private.

Figure 31. Percentage of Other DOD Component Initiatives Supporting Private or Public Sector Recipients, or Both.

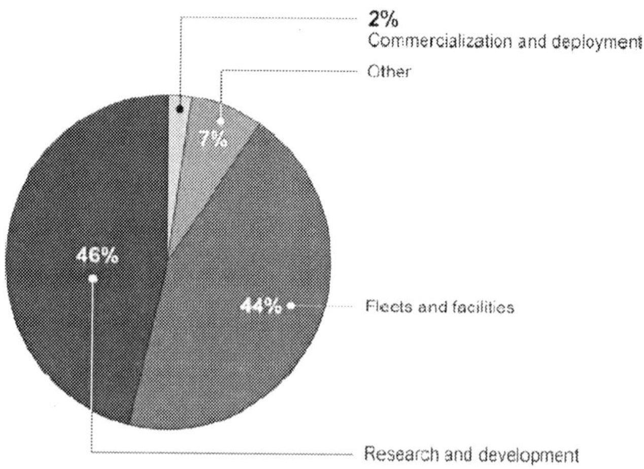

Source: GAO analysis of agency data.

Notes: This figure provides data on the federal roles for the 41 other DOD component renewable energy initiatives we identified. Percentages do not add to 100 due to rounding.

Figure 32. Percentage of Other DOD Component Initiatives by Federal Role and Other.

Department of Energy

The Department of Energy's (DOE) 92 initiatives supported every renewable energy source in our review, as demonstrated in figure 33. Most DOE initiatives supported more than a single energy source; among initiatives targeting a single energy source, solar and bioenergy were most commonly supported, as shown in figure 34. In addition, figure 35 demonstrates that more than 70 percent of DOE's initiatives supported both the public and private sectors. Figure 36 shows that DOE's initiatives were distributed across multiple federal roles, with the largest percentage of the agency's initiatives supporting research and development.

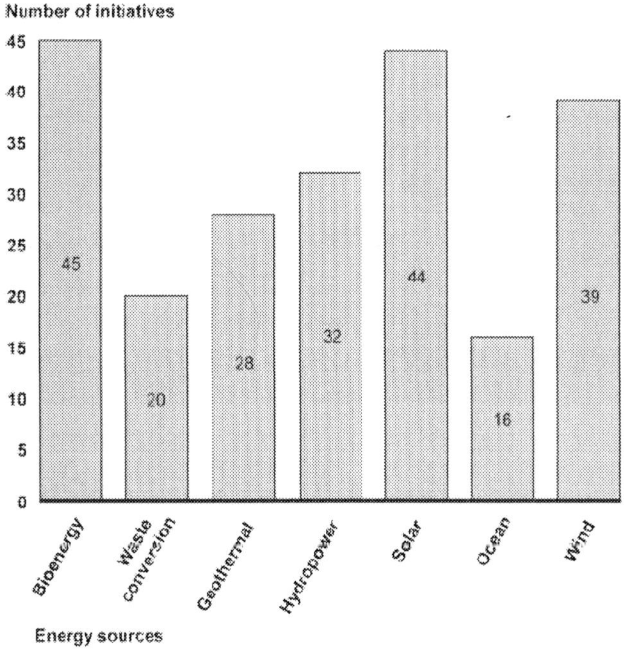

Source: GAO analysis of agency data.

Notes: The sum of the initiatives across the renewable energy sources adds to more than the total number of DOE initiatives because initiatives often supported multiple energy sources. Also, this figure does not include 15 initiatives for which we could not determine the applicable renewable energy sources based on the data DOE provided.

Figure 33. Number of DOE Initiatives Supporting Each Renewable Energy Source.

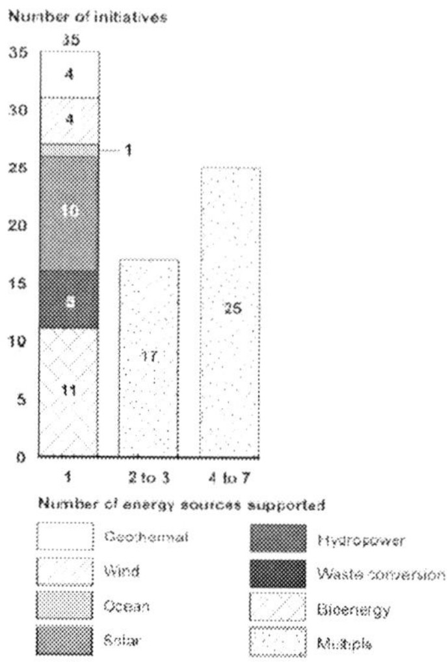

Source: GAO analysis of agency data.

Note: This figure does not include 15 initiatives for which we could not determine the applicable renewable energy sources based on the data DOE provided.

Figure 34. Number of Energy Sources Supported by DOE Initiatives.

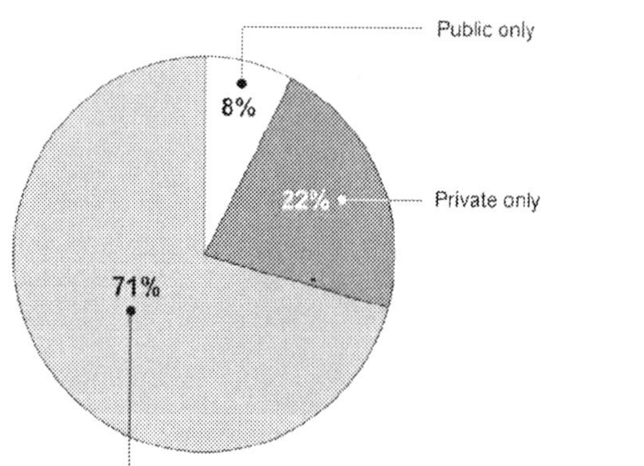

Source: GAO analysis of agency data.

Notes: This figure provides data on the recipients for the 92 DOE renewable energy initiatives we identified. We generally classified governmental recipients as public and nongovernmental recipients as private. Percentages do not add to 100 due to rounding.

Figure 35. Percentage of DOE Initiatives Supporting Private or Public Sector Recipients, or Both.

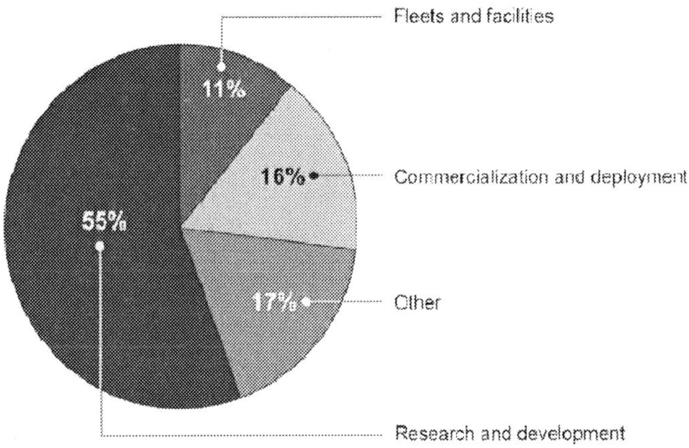

Source: GAO analysis of agency data.

Notes: This figure provides data on the federal roles for the 92 DOE renewable energy initiatives we identified. Percentages do not add to 100 due to rounding.

Figure 36. Percentage of DOE Initiatives by Federal Role and Other.

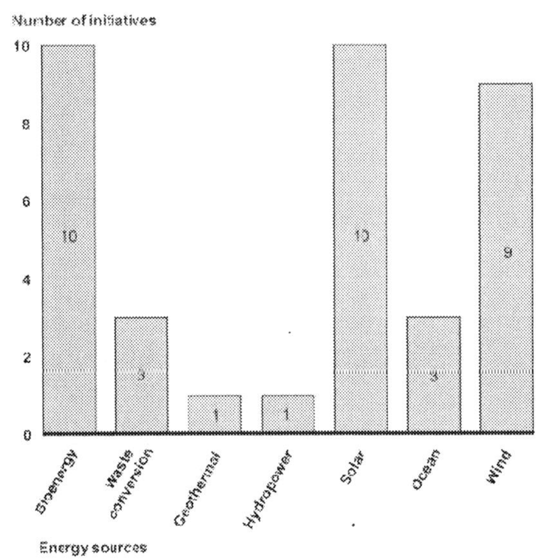

Source: GAO analysis of agency data.

Note: The sum of the initiatives across the renewable energy sources adds to more than the total number of DHS initiatives because initiatives often supported multiple energy sources.

Figure 37. Number of DHS Initiatives Supporting Each Renewable Energy Source.

Department of Homeland Security

The Department of Homeland Security's (DHS) 18 initiatives primarily supported three renewable energy sources—bioenergy, solar, and wind— as figure 37 shows.

DHS's initiatives were evenly split between those that supported more than one source and those that targeted a single source, with bioenergy and solar targeted the most by initiatives that supported a single energy source, as shown in figure 38.

Two-thirds of DHS's initiatives supported recipients in the private sector alone, as figure 39 indicates. Moreover, figure 40 shows that the majority of DHS's initiatives supported fleets and facilities.

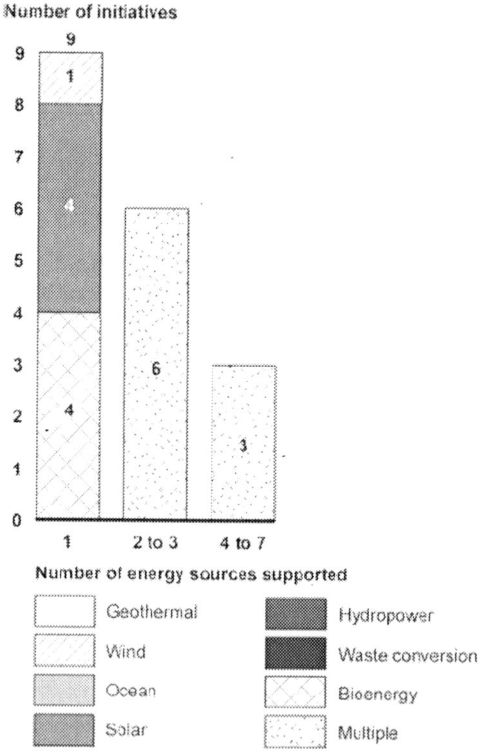

Source: GAO analysis of agency data.

Figure 38. Number of Energy Sources Supported by DHS Initiatives.

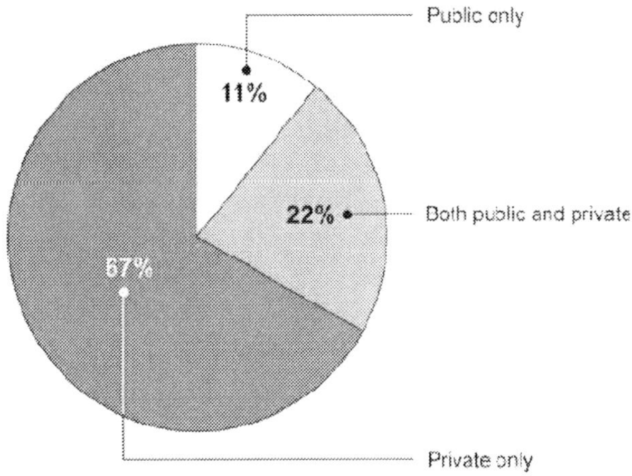

Source: GAO analysis of agency data.

Notes: This figure provides data on the recipients for the 18 DHS renewable energy initiatives we identified. We generally classified governmental recipients as public and nongovernmental recipients as private.

Figure 39. Percentage of DHS Initiatives Supporting Private or Public Sector Recipients, or Both.

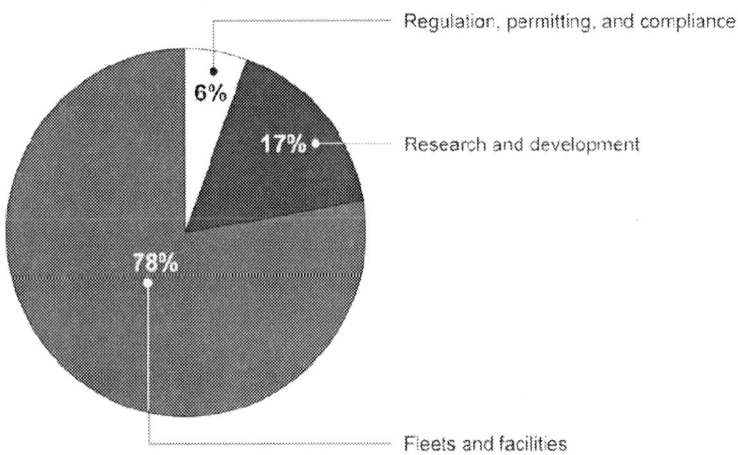

Source: GAO analysis of agency data.

Notes: This figure provides data on the federal roles for the 18 DHS renewable energy initiatives we identified. Percentages do not add to 100 due to rounding.

Figure 40. Percentage of DHS Initiatives by Federal Role and Other.

Department of Housing and Urban Development

The Department of Housing and Urban Development's (HUD) 13 initiatives supported 3 of the 7 energy sources in our review, as figure 41 shows. In addition, figure 42 shows that most HUD initiatives supported multiple energy sources. Figure 43 demonstrates that most of HUD's initiatives supported either private or public sector recipients with fewer initiatives supporting both. Three-fourths of HUD's initiatives supported commercialization and deployment, with the remainder supporting fleet and facilities efforts, as shown in figure 44.

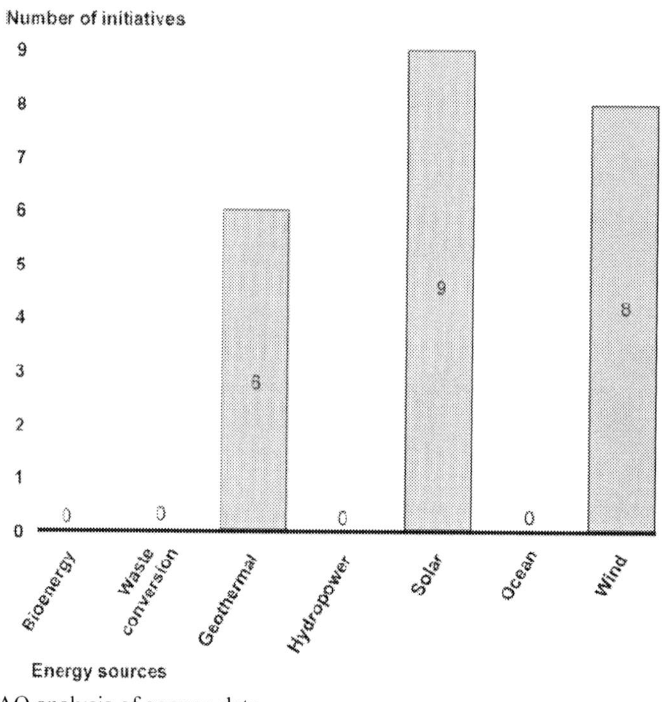

Source: GAO analysis of agency data.

Notes: The sum of the initiatives across the renewable energy sources adds to more than the total number of HUD initiatives because initiatives often supported multiple energy sources. Also, this figure does not include four initiatives for which we could not determine the applicable renewable energy sources based on the data HUD provided.

Figure 41. Number of HUD Initiatives Supporting Each Renewable Energy Source.

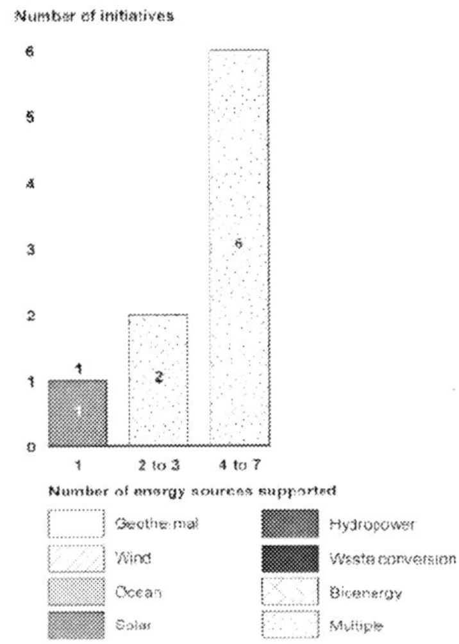

Source: GAO analysis of agency data.

Note: This figure does not include four initiatives for which we could not determine the applicable renewable energy sources based on the data HUD provided.

Figure 42. Number of Energy Sources Supported by HUD Initiatives.

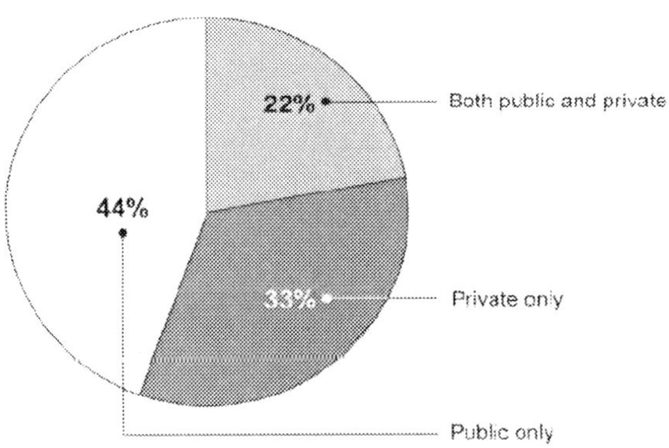

Source: GAO analysis of agency data.

Notes: This figure provides data on the recipients for 9 of the 13 HUD renewable energy initiatives we identified. The figure does not include four initiatives for which we could not determine the applicable recipients based on the data HUD provided. We generally classified governmental recipients as public and nongovernmental recipients as private. Percentages do not add to 100 due to rounding.

Figure 43. Percentage of HUD Initiatives Supporting Private or Public Sector Recipients, or Both.

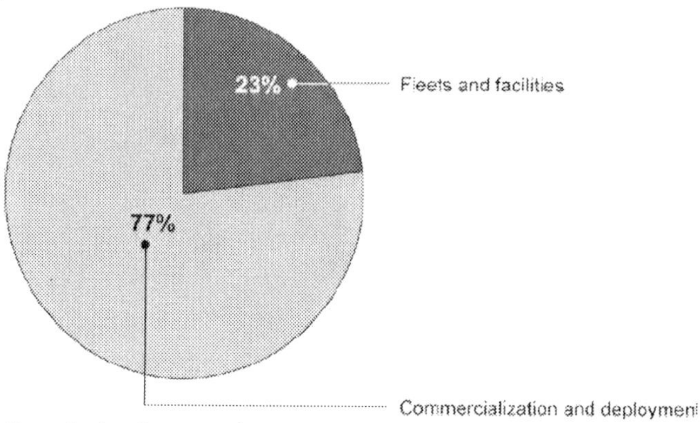

Source: GAO analysis of agency data.

Note: This figure provides data on the federal roles for the 13 HUD renewable energy initiatives we identified.

Figure 44. Percentage of HUD Initiatives by Federal Role and Other.

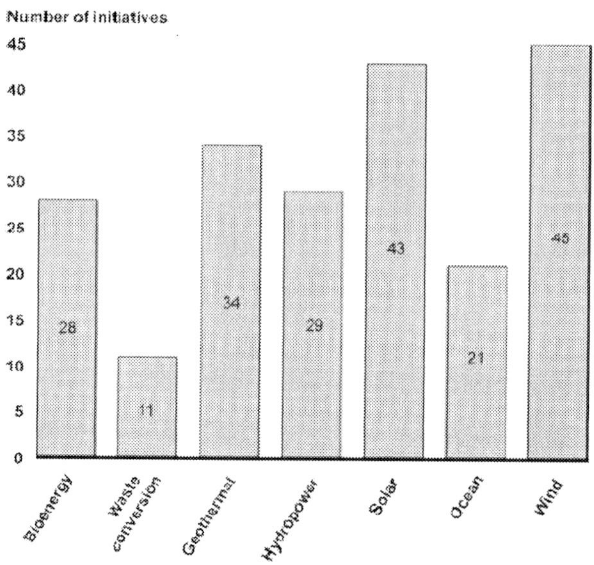

Source: GAO analysis of agency data.

Note: The sum of the initiatives across the renewable energy sources adds to more than the total number of Interior initiatives because initiatives often supported multiple energy sources.

Figure 45. Number of Interior Initiatives Supporting Each Renewable Energy Source.

Department of the Interior

The Department of the Interior's (Interior) 82 initiatives supported all of the renewable energy sources in our review, with wind and solar power supported the most, as figure 45 shows. Nearly half of Interior's initiatives were targeted to a single energy source, although that energy source varied, as figure 46 demonstrates. Figure 47 shows that more than half of Interior's initiatives supported both private and public sector recipients, and figure 48 indicates that the initiatives were distributed across all of the key federal roles, with more than half supporting the roles of regulation, permitting, and compliance and fleets and facilities.

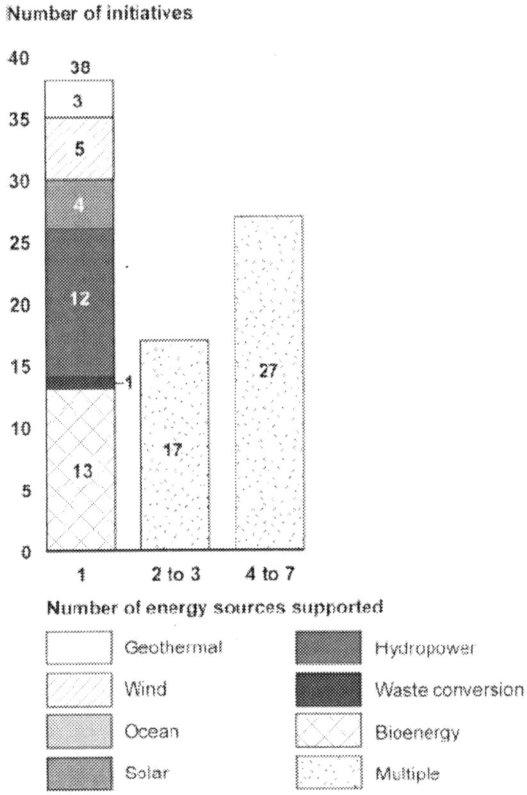

Source: GAO analysis of agency data.

Figure 46. Number of Energy Sources Supported by Interior Initiatives.

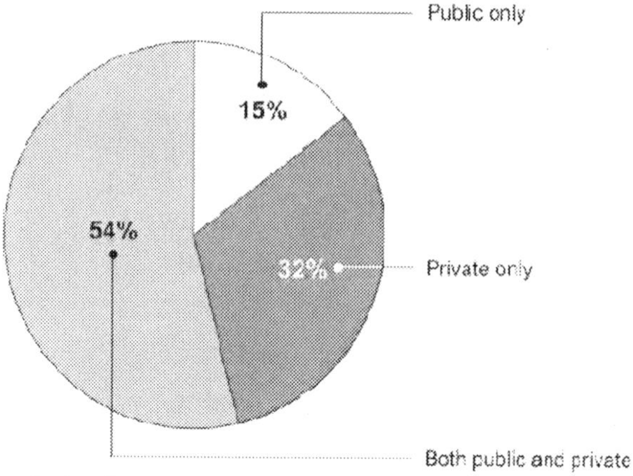

Source: GAO analysis of agency data.

Notes: This figure provides data on the recipients for the 82 Interior renewable energy initiatives we identified. We generally classified governmental recipients as public and nongovernmental recipients as private. Percentages do not add to 100 due to rounding.

Figure 47. Percentage of Interior Initiatives Supporting Private or Public Sector Recipients, or Both.

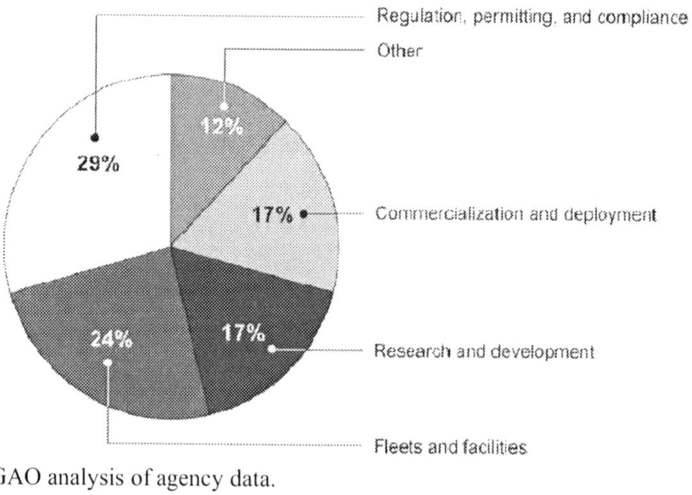

Source: GAO analysis of agency data.

Notes: This figure provides data on the federal roles for the 82 Interior renewable energy initiatives we identified. Percentages do not add to 100 due to rounding.

Figure 48. Percentage of Interior Initiatives by Federal Role and Other.

Department of Justice

The Department of Justice's (Justice) 14 initiatives supported 5 of the 7 energy sources in our review, with bioenergy receiving the most support and targeted most often, as figures 49 and 50 show, respectively. In addition, figures 51 and 52, respectively, demonstrate that almost all of Justice's initiatives supported private sector recipients alone and the federal role of fleets and facilities.

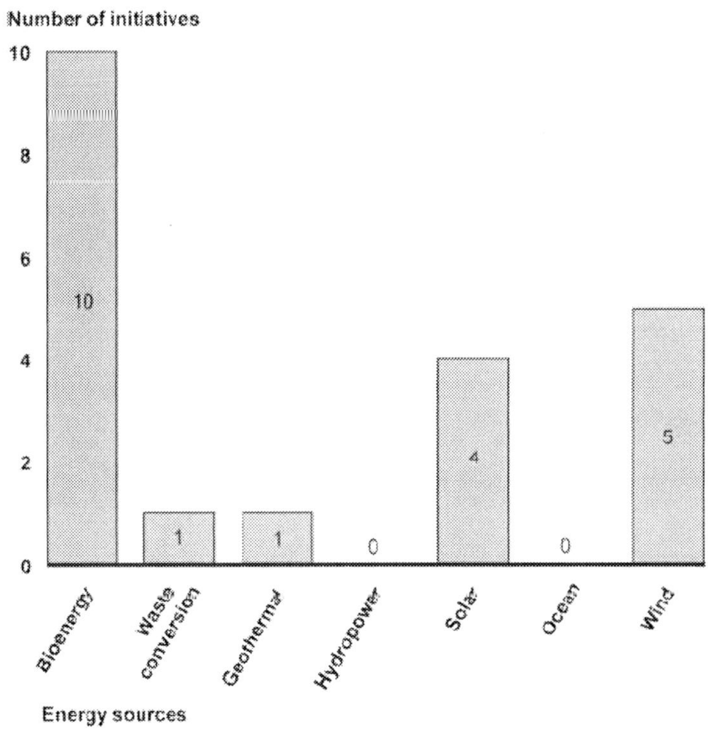

Source: GAO analysis of agency data.

Note: The sum of the initiatives across the renewable energy sources adds to more than the total number of Justice initiatives because some initiatives supported multiple energy sources.

Figure 49. Number of Justice Initiatives Supporting Each Renewable Energy Source.

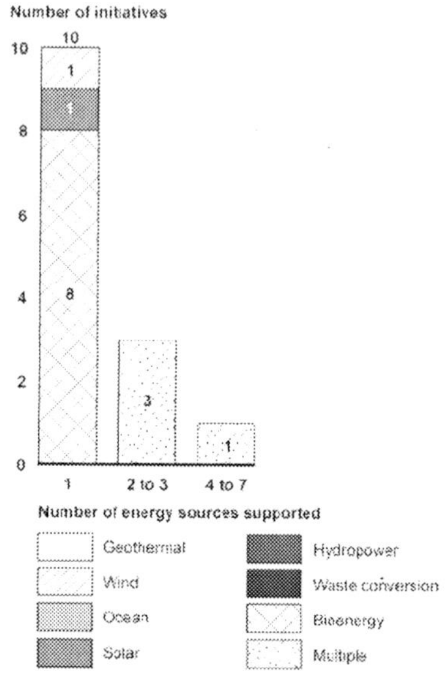

Source: GAO analysis of agency data.

Figure 50. Number of Energy Sources Supported by Justice Initiatives.

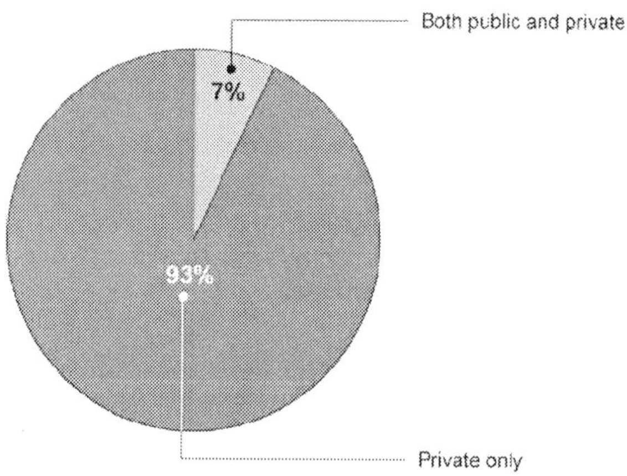

Source: GAO analysis of agency data.

Notes: This figure provides data on the recipients for the 14 Justice renewable energy initiatives we identified. We generally classified governmental recipients as public and nongovernmental recipients as private.

Figure 51. Percentage of Justice Initiatives Supporting Private or Public Sector Recipients, or Both.

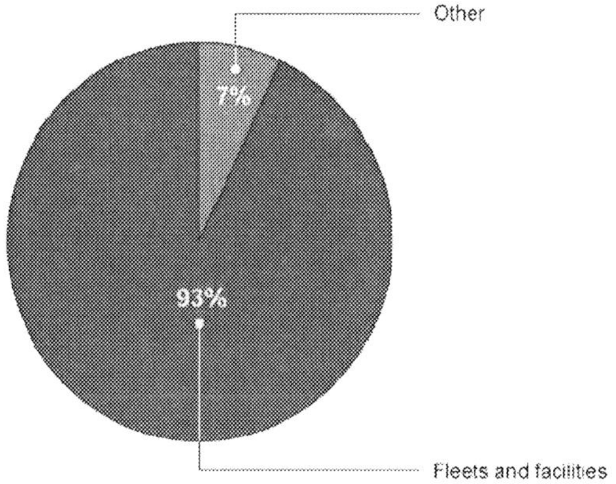

Source: GAO analysis of agency data.

Note: This figure provides data on the federal roles for the 14 Justice renewable energy initiatives we identified.

Figure 52. Percentage of Justice Initiatives by Federal Role and Other.

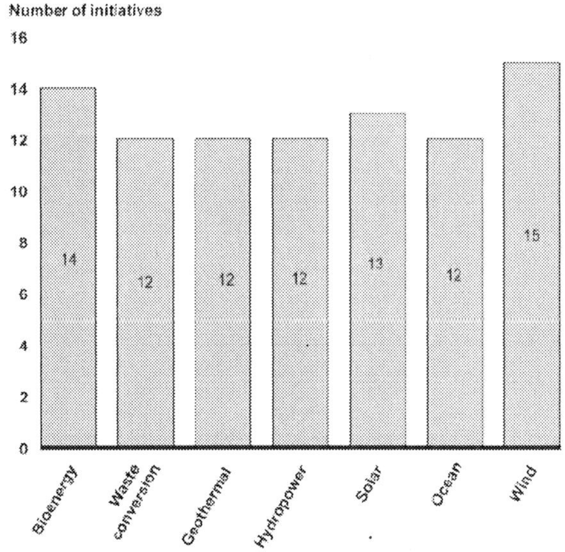

Source: GAO analysis of agency data.

Note: The sum of the initiatives across the renewable energy sources adds to more than the total number of Labor initiatives because initiatives often supported multiple energy sources.

Figure 53. Number of Labor Initiatives Supporting Each Renewable Energy Source.

Department of Labor

The Department of Labor's (Labor) 16 initiatives were fairly evenly distributed in their support of energy sources and most initiatives supported all 7 energy sources in our review, as shown in figures 53 and 54, respectively. Figure 55 shows that more than two-thirds of Labor's initiatives supported recipients in the private sector, and more than 80 percent of them supported efforts outside of the four key federal roles in our review, with the rest supporting fleets and facilities, as shown in figure 56.

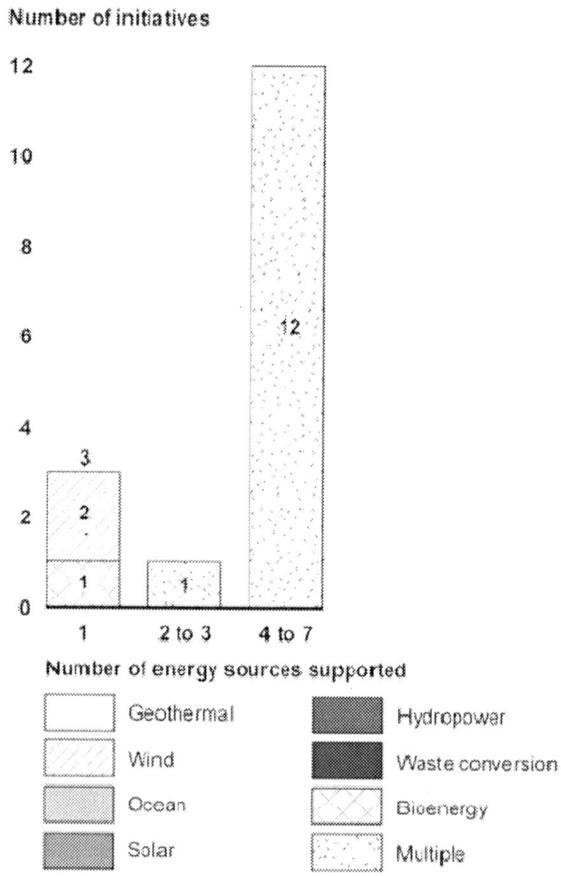

Source: GAO analysis of agency data.

Figure 54. Number of Energy Sources Supported by Labor Initiatives.

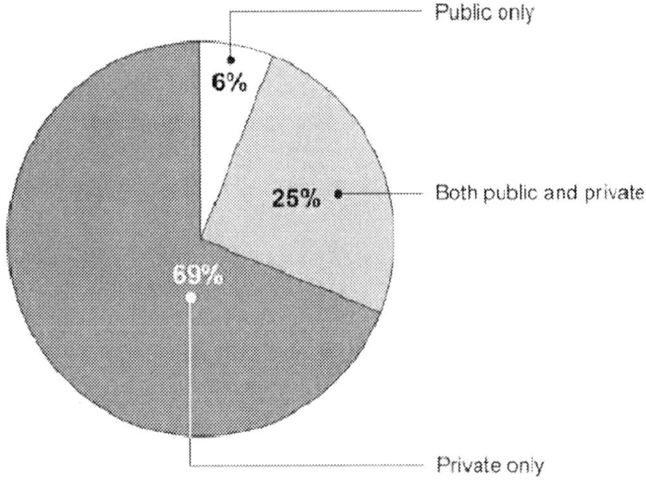

Source: GAO analysis of agency data.

Notes: This figure provides data on the recipients for the 16 Labor renewable energy initiatives we identified. We generally classified governmental recipients as public and nongovernmental recipients as private.

Figure 55. Percentage of Labor Initiatives Supporting Private or Public Sector Recipients, or Both.

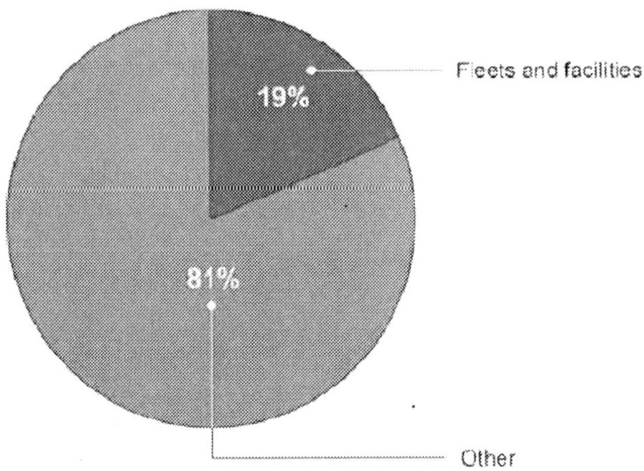

Source: GAO analysis of agency data.

Note: This figure provides data on the federal roles for the 16 Labor renewable energy initiatives we identified.

Figure 56. Percentage of Labor Initiatives by Federal Role and Other.

Department of State

The Department of State's (State) 20 initiatives provided support for all of the energy sources in our review, as figure 57 demonstrates. Figure 58 shows that all of State's five targeted initiatives supported bioenergy, with the rest of the initiatives supporting multiple sources. Figure 59 shows that State's initiatives supported a mix of public and private sector recipients or both, while figure 60 demonstrates that 60 percent of State's initiatives supported efforts outside of the four key federal roles in our analysis, such as training and capacity building.

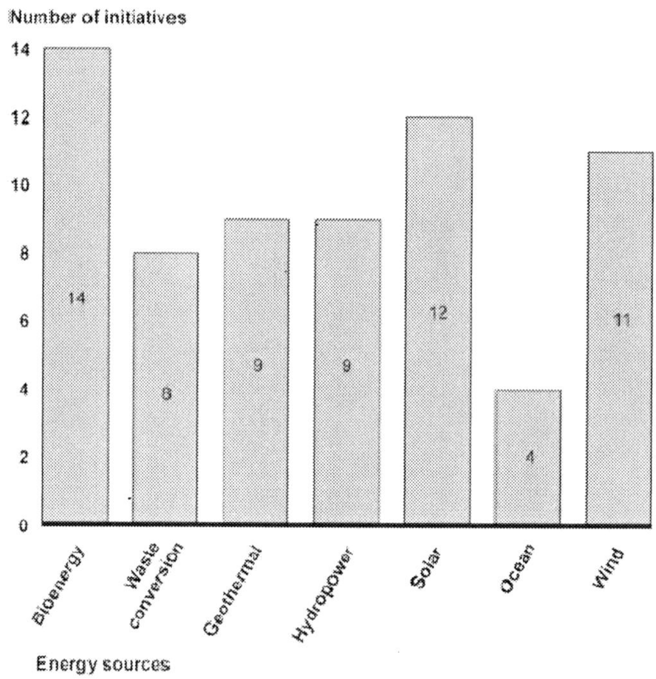

Source: GAO analysis of agency data.

Notes: The sum of the initiatives across the renewable energy sources adds to more than the total number of State initiatives because initiatives often supported multiple energy sources. Also, this figure does not include one initiative for which we could not determine the applicable renewable energy sources based on the data State provided.

Figure 57. Number of State Initiatives Supporting Each Renewable Energy Source.

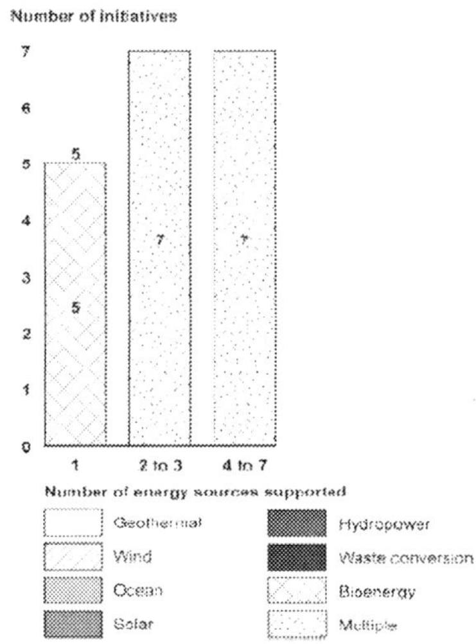

Source: GAO analysis of agency data.

Note: This figure does not include one initiative for which we could not determine the applicable renewable energy sources based on the data State provided.

Figure 58. Number of Energy Sources Supported by State Initiatives.

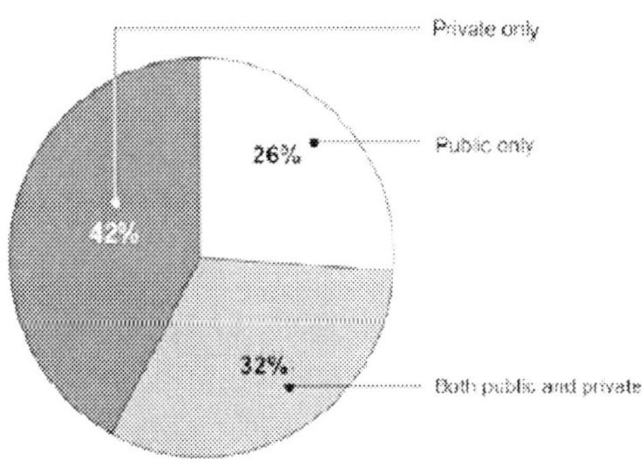

Source: GAO analysis of agency data.

Notes: This figure provides data on the recipients for 19 of the 20 State renewable energy initiatives we identified. The figure does not include one initiative for which we could not determine the applicable recipients based on the data State provided. We generally classified governmental recipients as public and nongovernmental recipients as private.

Figure 59. Percentage of State Initiatives Supporting Private or Public Sector Recipients, or Both.

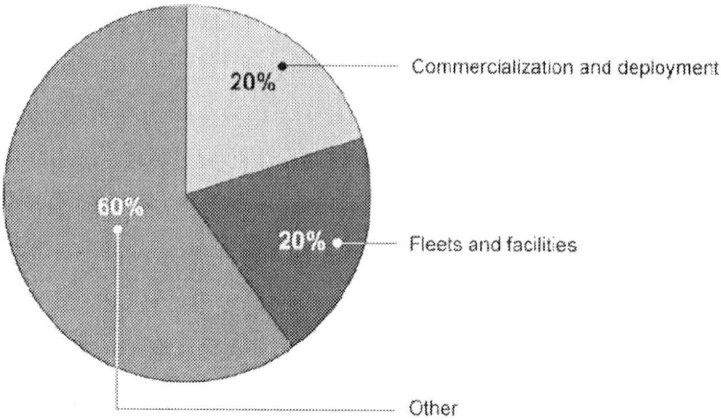

Source: GAO analysis of agency data.

Notes: This figure provides data on the federal roles for the 20 State renewable energy initiatives we identified.

Figure 60. Percentage of State Initiatives by Federal Role and Other.

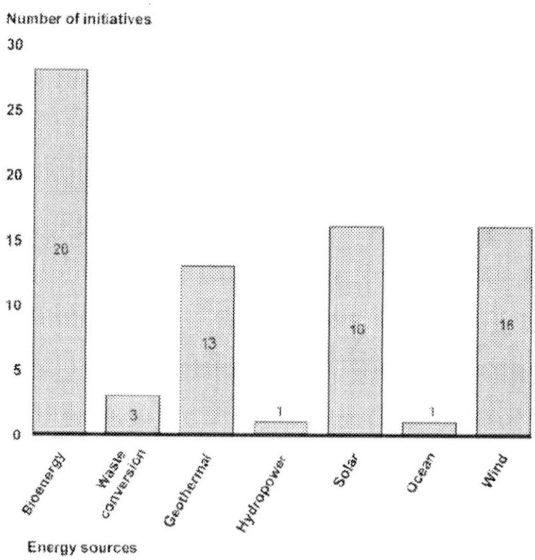

Source: GAO analysis of agency data.

Note: The sum of the initiatives across the renewable energy sources adds to more than the total number of DOT initiatives because initiatives often supported multiple energy sources.

Figure 61. Number of DOT Initiatives Supporting Each Renewable Energy Source.

Department of Transportation

The Department of Transportation's (DOT) 36 initiatives supported all of the initiatives in our review, with bioenergy supported most often, as shown in figure 61. In addition, almost all of the 17 targeted initiatives supported bioenergy, although a majority of DOT's initiatives supported more than one energy source, as demonstrated in figure 62. Initiatives also supported both public and private sector recipients, as figure 63 shows, with private sector recipients alone supported by 50 percent of the initiatives. In addition, figure 64 demonstrates that DOT's initiatives were distributed across all four of the main federal roles in our review.

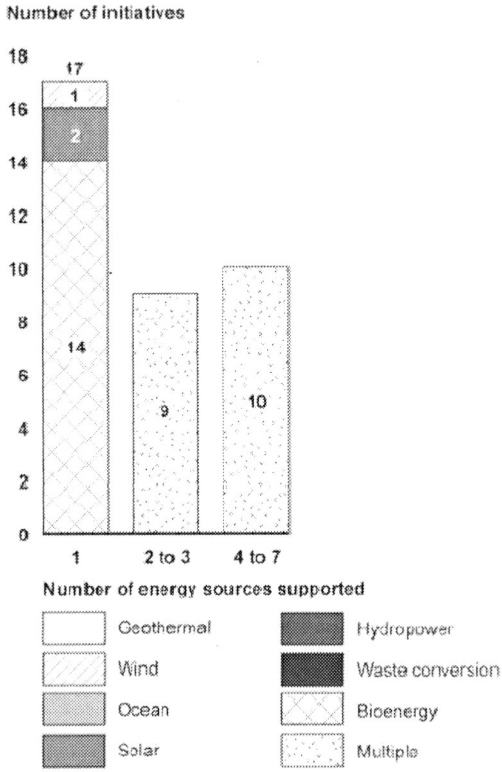

Source: GAO analysis of agency data.

Figure 62. Number of Energy Sources Supported by DOT Initiatives.

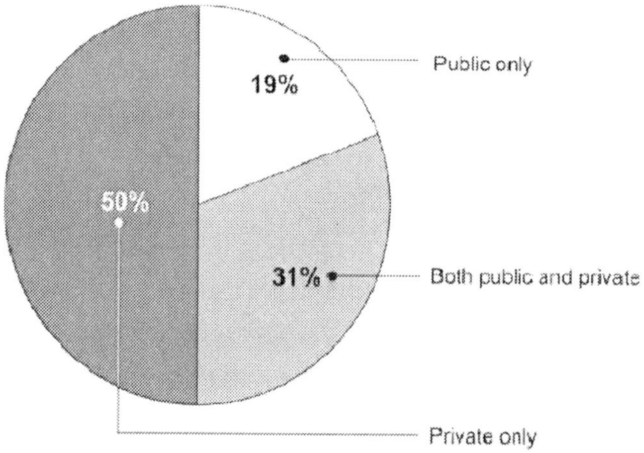

Source: GAO analysis of agency data.

Notes: This figure provides data on the recipients for the 36 DOT renewable energy initiatives we identified. We generally classified governmental recipients as public and nongovernmental recipients as private.

Figure 63. Percentage of DOT Initiatives Supporting Private or Public Sector Recipients, or Both.

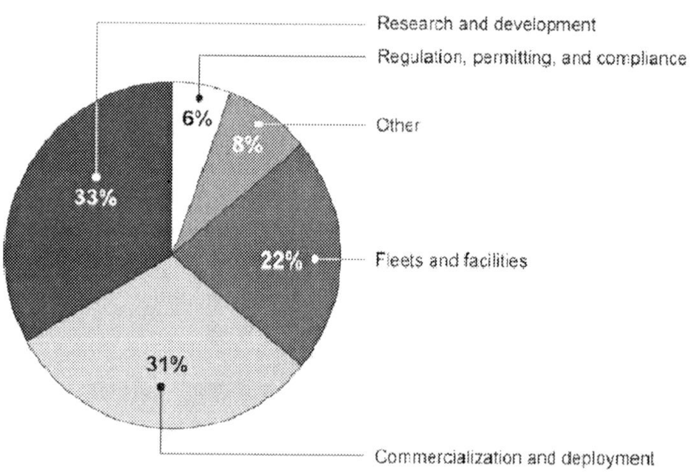

Source: GAO analysis of agency data.

Note: This figure provides data on the federal roles for the 36 DOT renewable energy initiatives we identified.

Figure 64. Percentage of DOT Initiatives by Federal Role and Other.

Department of the Treasury

The Department of the Treasury's (Treasury) 31 initiatives supported all the energy sources in our review, with bioenergy most commonly supported by initiatives and more than a third of initiatives targeted bioenergy exclusively, as shown in figures 65 and 66, respectively.

Figure 67 shows that two-thirds of Treasury's initiatives supported private sector recipients alone. Nearly three-fourths of Treasury's initiatives supported the federal role of commercialization and deployment, as demonstrated in figure 68.

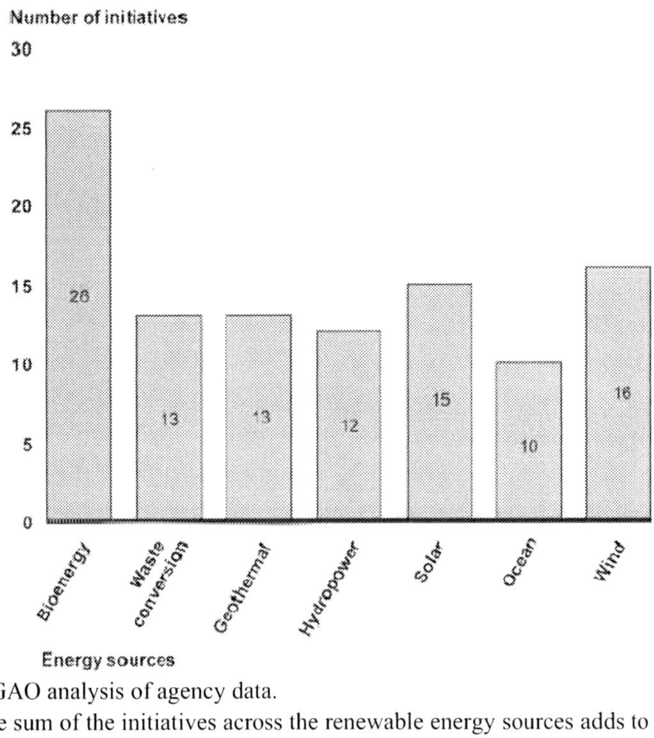

Source: GAO analysis of agency data.

Note: The sum of the initiatives across the renewable energy sources adds to more than the total number of Treasury initiatives because initiatives often supported multiple energy sources.

Figure 65. Number of Treasury Initiatives Supporting Each Renewable Energy Source.

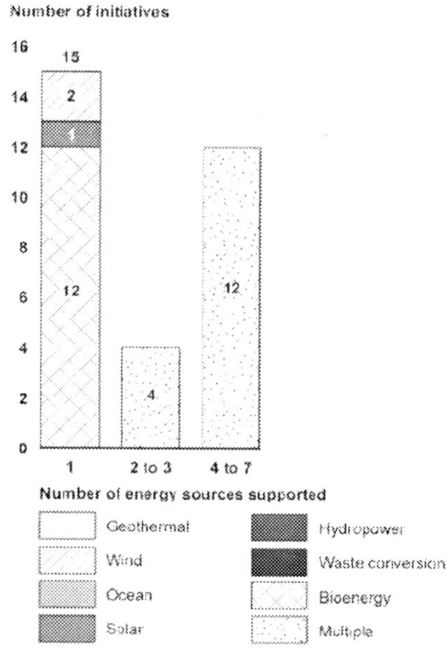

Source: GAO analysis of agency data.

Figure 66. Number of Energy Sources Supported by Treasury Initiatives.

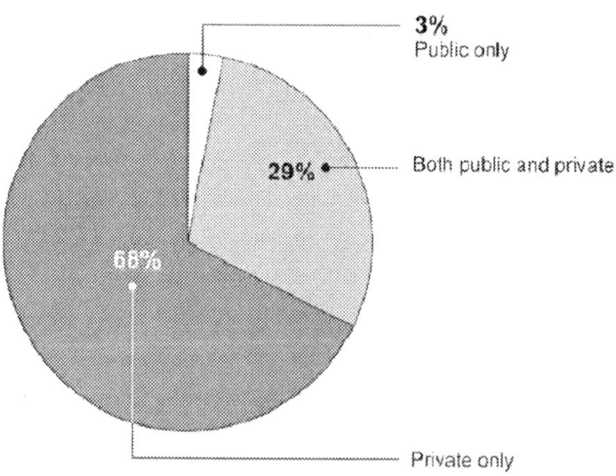

Source: GAO analysis of agency data.

Notes: This figure provides data on the recipients for the 31 Treasury renewable energy initiatives we identified. We generally classified governmental recipients as public and nongovernmental recipients as private.

Figure 67. Percentage of Treasury Initiatives Supporting Private or Public Sector Recipients, or Both.

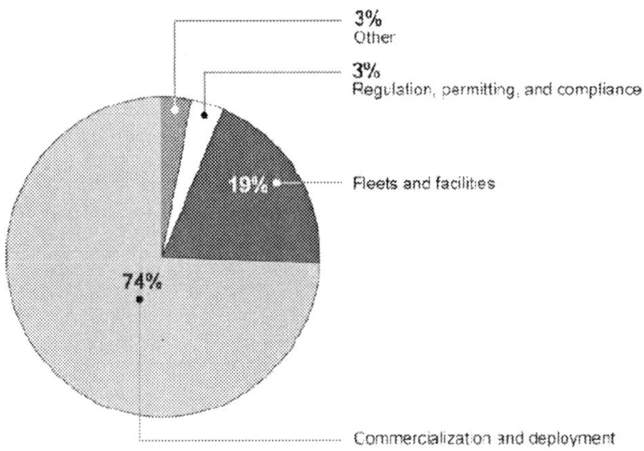

Source: GAO analysis of agency data.

Notes: This figure provides data on the federal roles for the 31 Treasury renewable energy initiatives we identified. Percentages do not add to 100 due to rounding.

Figure 68. Percentage of Treasury Initiatives by Federal Role and Other.

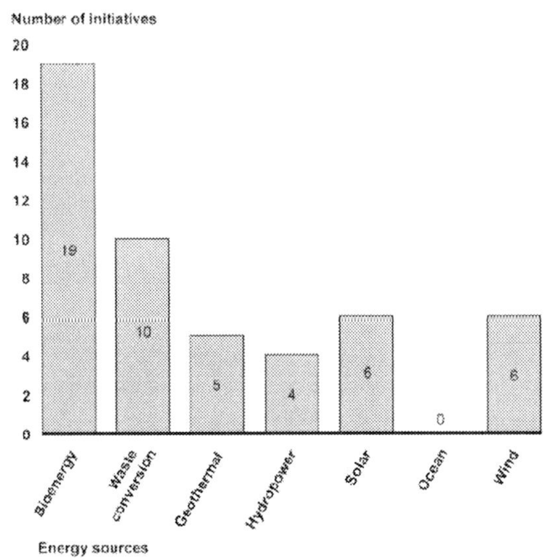

Source: GAO analysis of agency data.

Note: The sum of the initiatives across the renewable energy sources adds to more than the total number of EPA initiatives because initiatives often supported multiple energy sources.

Figure 69. Number of EPA Initiatives Supporting Each Renewable Energy Source.

Environmental Protection Agency

Figure 69 demonstrates that most of the Environmental Protection Agency's (EPA) 22 initiatives supported bioenergy, while a smaller number supported all other energy sources in our review except for ocean energy. Half of EPA's initiatives supported a single energy source, with bioenergy the most targeted source, as figure 70 shows. Figure 71 demonstrates that half of EPA's initiatives supported recipients in the private sector alone, with most of the rest supporting both public and private sector recipients. Figure 72 shows that EPA's initiatives supported all four key federal roles.

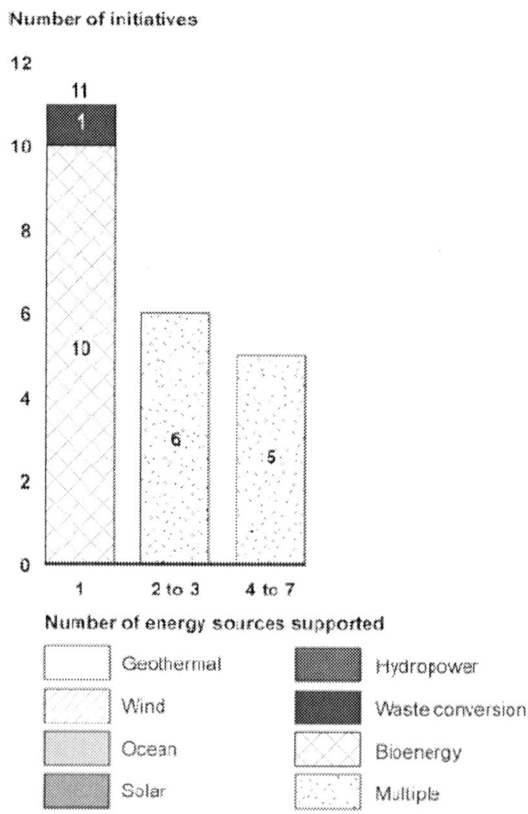

Source: GAO analysis of agency data.

Figure 70. Number of Energy Sources Supported by EPA Initiatives.

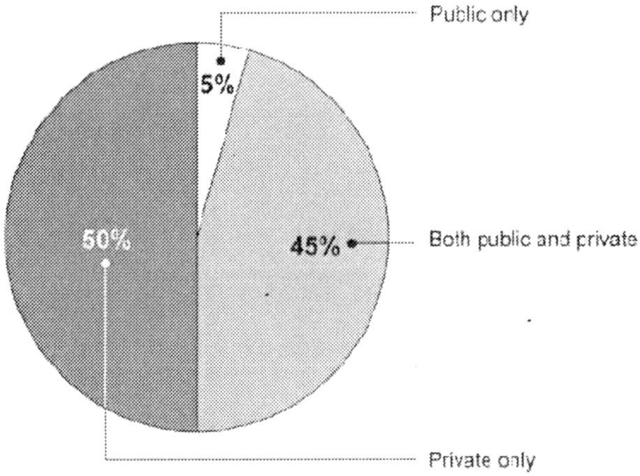

Source: GAO analysis of agency data.

Notes: This figure provides data on the recipients for the 22 EPA renewable energy initiatives we identified. We generally classified governmental recipients as public and nongovernmental recipients as private.

Figure 71. Percentage of EPA Initiatives Supporting Private or Public Sector Recipients, or Both.

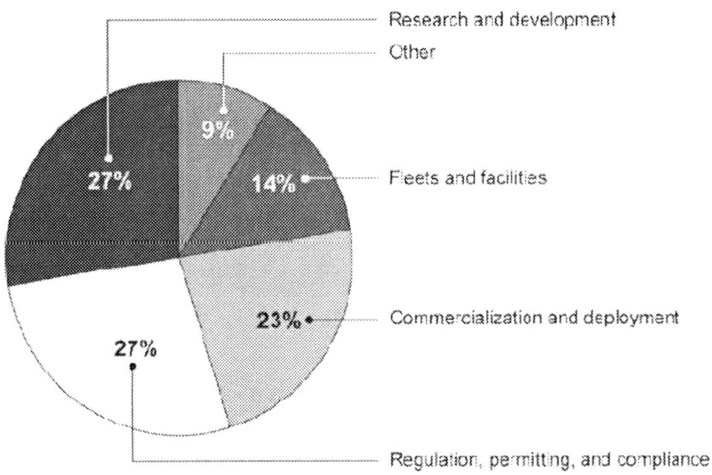

Source: GAO analysis of agency data.

Note: This figure provides data on the federal roles for the 22 EPA renewable energy initiatives we identified.

Figure 72. Percentage of EPA Initiatives by Federal Role and Other.

Federal Energy Regulatory Commission

The Federal Energy Regulatory Commission's (FERC) seven initiatives provided support for five of the seven energy sources with wind and solar power supported the most, as figure 73 shows. Three FERC initiatives each targeted a single source, shown in figure 74, while the other FERC initiatives supported between two and three energy sources each. More than half of FERC's initiatives supported private sector recipients alone, as shown in figure 75, while figure 76 shows more than 40 percent of initiatives supported the regulation, permitting, and compliance federal role with the rest evenly split between research and development and fleets and facilities.

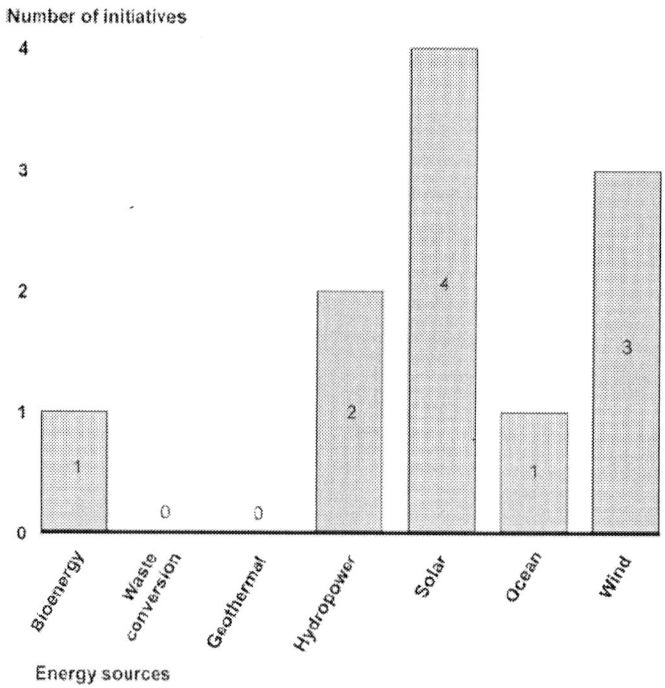

Source: GAO analysis of agency data.

Note: The sum of the initiatives across the renewable energy sources adds to more than the total number of FERC initiatives because initiatives often supported multiple energy sources.

Figure 73. Number of FERC Initiatives Supporting Each Renewable Energy Source.

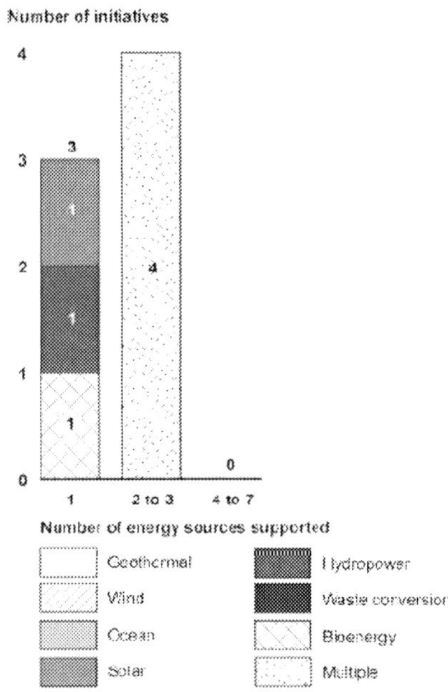

Source: GAO analysis of agency data.

Figure 74. Number of Energy Sources Supported by FERC Initiatives.

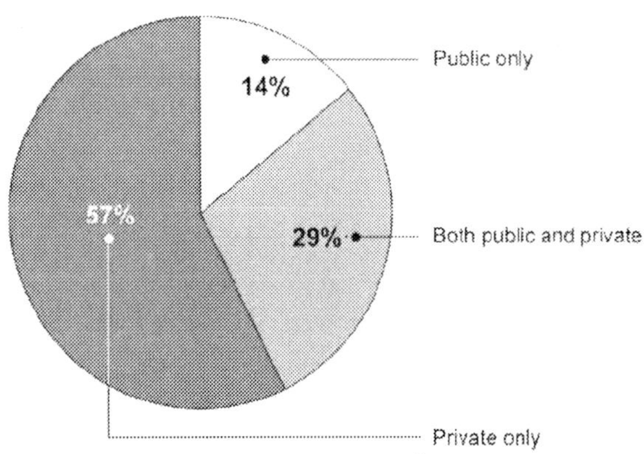

Source: GAO analysis of agency data.

Notes: This figure provides data on the recipients for the seven FERC renewable energy initiatives we identified. We generally classified governmental recipients as public and nongovernmental recipients as private.

Figure 75. Percentage of FERC Initiatives Supporting Private or Public Sector Recipients, or Both.

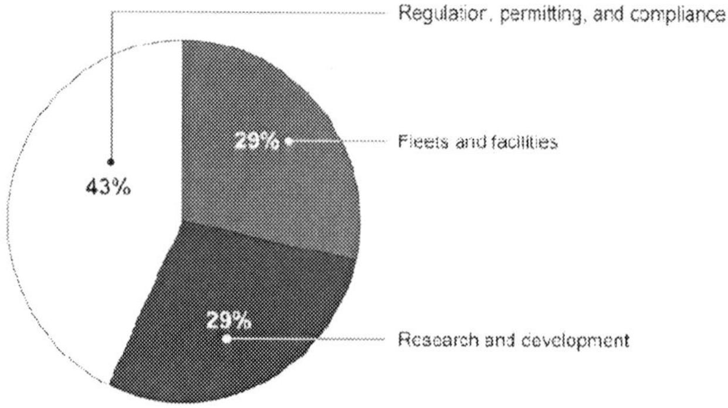

Source: GAO analysis of agency data.

Notes: This figure provides data on the federal roles for the seven FERC renewable energy initiatives we identified. Percentages do not add to 100 due to rounding.

Figure 76. Percentage of FERC Initiatives by Federal Role and Other.

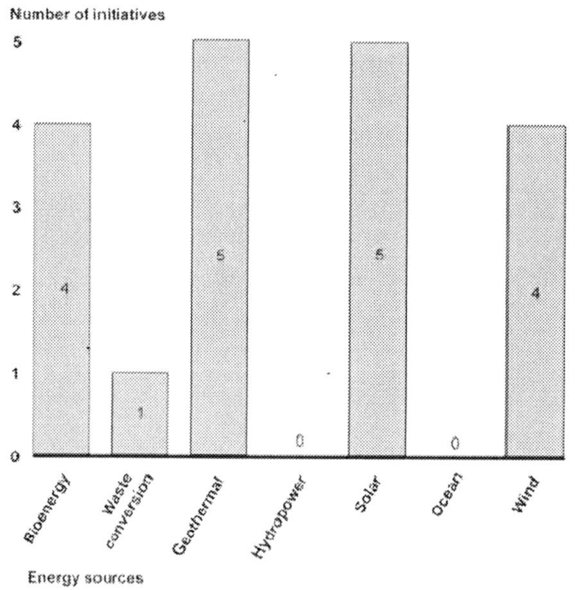

Source: GAO analysis of agency data.

Note: The sum of the initiatives across the renewable energy sources adds to more than the total number of GSA initiatives because initiatives often supported multiple energy sources.

Figure 77. Number of GSA Initiatives Supporting Each Renewable Energy Source.

General Services Administration

The General Services Administration's (GSA) nine initiatives were generally evenly distributed across the five energy sources that they supported with waste conversion supported less commonly, as shown in figure 77. Of the four initiatives that targeted a single source, figure 78 demonstrates that three supported bioenergy. Figure 79 shows that more than half of GSA's initiatives supported private sector recipients alone. In addition, figure 80 demonstrates that most GSA initiatives supported fleets and facilities, with the rest supporting research and development.

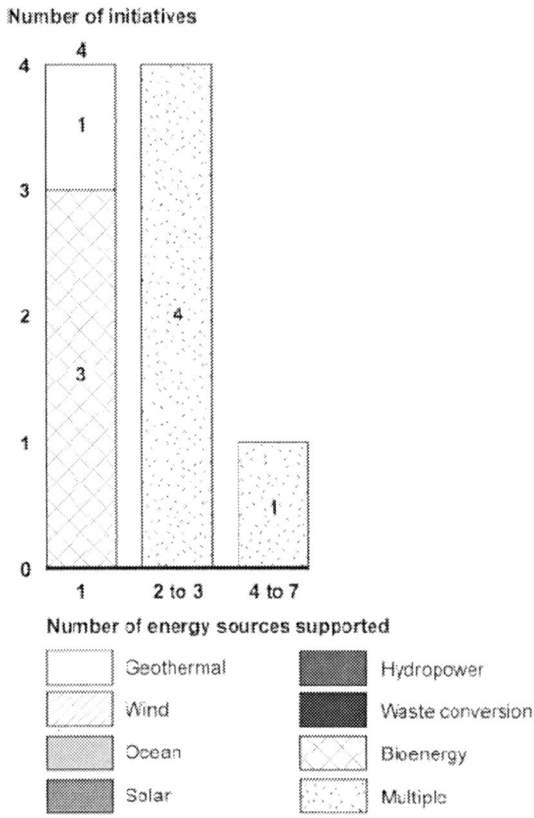

Source: GAO analysis of agency data.

Figure 78. Number of Energy Sources Supported by GSA Initiatives.

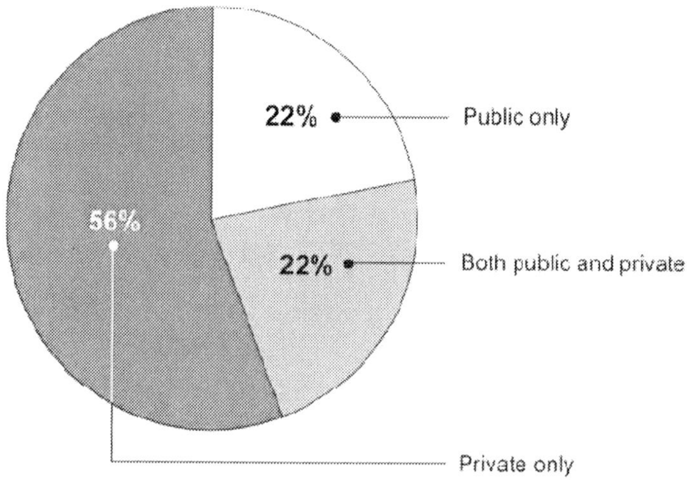

Source: GAO analysis of agency data.

Notes: This figure provides data on the recipients for the nine GSA renewable energy initiatives we identified. We generally classified governmental recipients as public and nongovernmental recipients as private.

Figure 79. Percentage of GSA Initiatives Supporting Private or Public Sector Recipients, or Both.

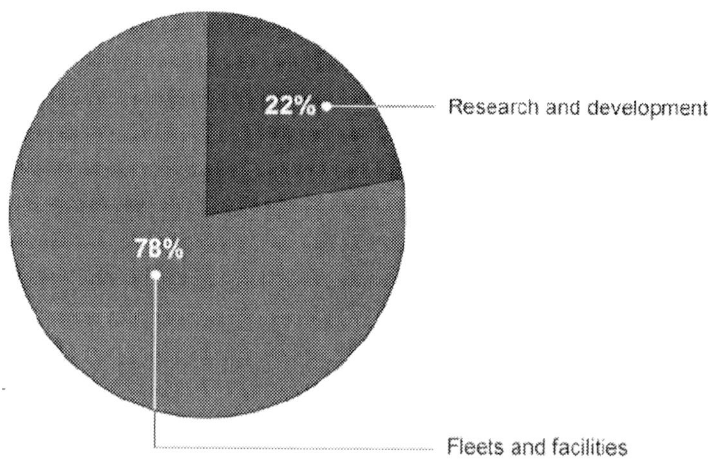

Source: GAO analysis of agency data.

Note: This figure provides data on the federal roles for the nine GSA renewable energy initiatives we identified.

Figure 80. Percentage of GSA Initiatives by Federal Role and Other.

National Aeronautics and Space Administration

Figure 81 shows that the National Aeronautics and Space Administration's (NASA) 12 initiatives supported 5 of the 7 energy sources in our review, with solar power supported most commonly. Three of NASA's four targeted initiatives provided support for solar power, as figure 82 shows. In addition, figure 83 shows that two-thirds of NASA's initiatives supported private sector recipients alone. More than 60 percent of NASA's initiatives supported the fleets and facilities federal role, with the rest supporting research and development, as demonstrated in figure 84.

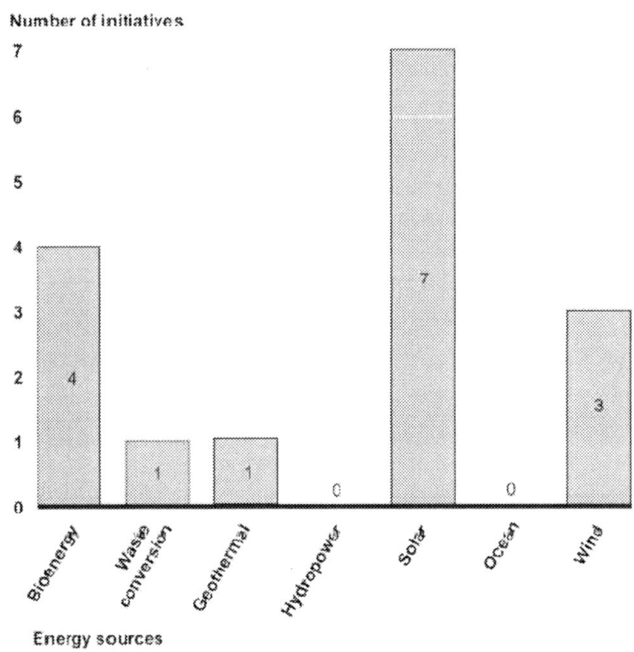

Source: GAO analysis of agency data.

Notes: The sum of the initiatives across the renewable energy sources adds to more than the total number of NASA initiatives because initiatives often supported multiple energy sources. Also, this figure does not include three initiatives for which we could not determine the applicable renewable energy sources based on the data NASA provided.

Figure 81. Number of NASA Initiatives Supporting Each Renewable Energy Source.

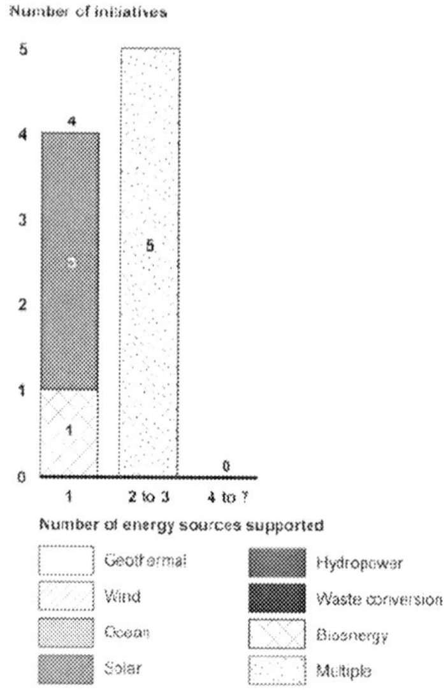

Source: GAO analysis of agency data.

Note: This figure does not include three initiatives for which we could not determine the applicable renewable energy sources based on the data NASA provided.

Figure 82. Number of Energy Sources Supported by NASA Initiatives.

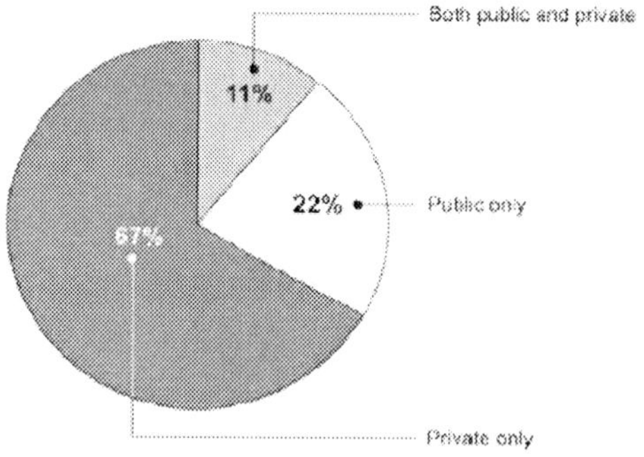

Source: GAO analysis of agency data.

Notes: This figure provides data on the recipients for 9 of the 12 NASA renewable energy initiatives we identified. The figure does not include three initiatives for which we could not determine the applicable recipients based on the data NASA provided. We generally classified governmental recipients as public and nongovernmental recipients as private.

Figure 83. Percentage of NASA Initiatives Supporting Private or Public Sector Recipients, or Both.

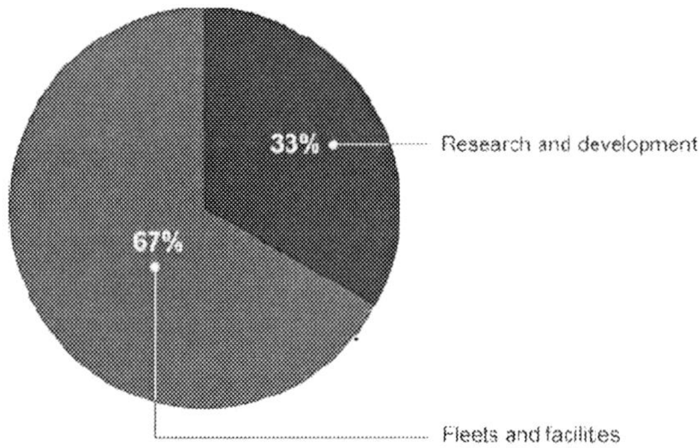

Source: GAO analysis of agency data.

Notes: This figure provides data on the federal roles for 9 of the 12 NASA renewable energy initiatives we identified. The figure does not include three initiatives for which we could not determine the applicable federal role based on the data NASA provided.

Figure 84. Percentage of NASA Initiatives by Federal Role and Other.

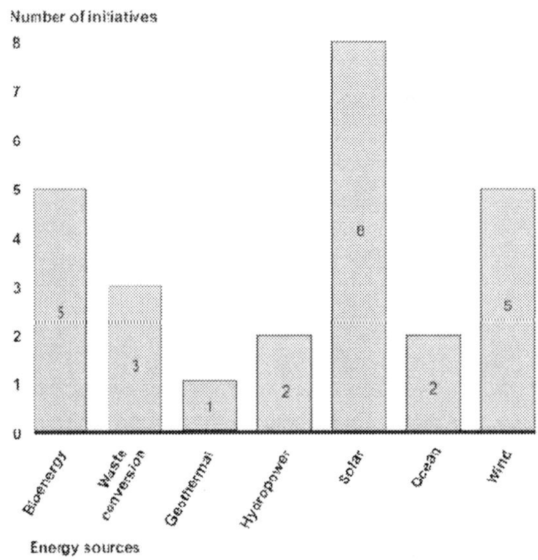

Source: GAO analysis of agency data.

Note: The sum of the initiatives across the renewable energy sources adds to more than the total number of NSF initiatives because initiatives often supported multiple energy sources.

Figure 85. Number of NSF Initiatives Supporting Each Renewable Energy Source.

National Science Foundation

Figure 85 shows that the National Science Foundation's (NSF) nine initiatives supported all of the energy sources in our review with solar, wind, and bioenergy receiving the most support. Most initiatives supported more than one energy source, as figure 86 demonstrates. In addition, figures 87 and 88, respectively, show that all of the initiatives supported only private sector recipients, with almost 90 percent supporting the federal role of research and development.[2]

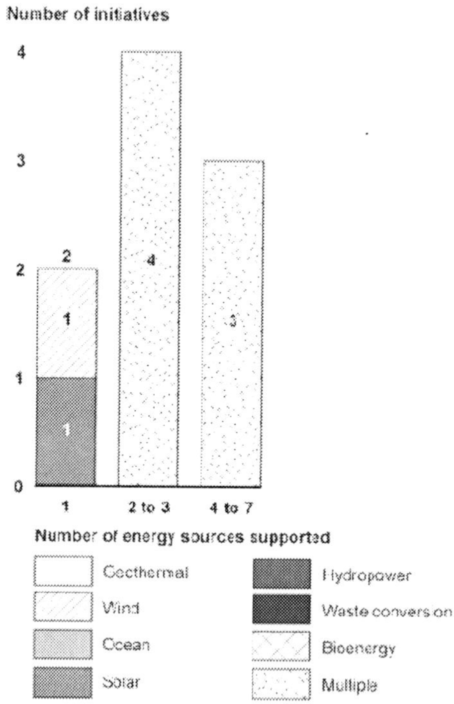

Source: GAO analysis of agency data.

Notes: This figure provides data on the recipients for the nine NSF renewable energy initiatives we identified. We generally classified governmental recipients as public and nongovernmental recipients as private.

Figure 86. Number of Energy Sources Supported by NSF Initiatives.

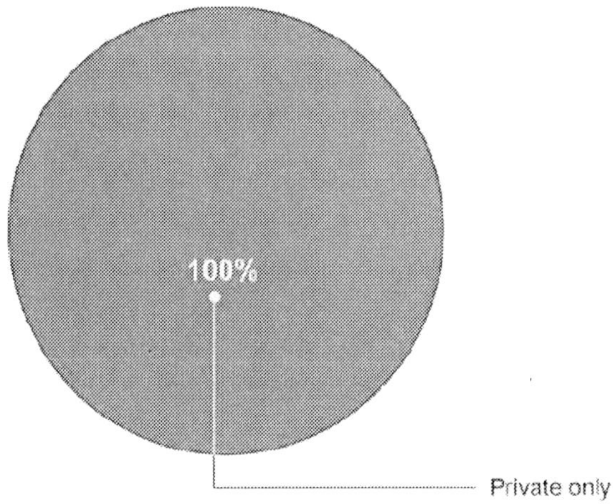

Source: GAO analysis of agency data.

Notes: This figure provides data on the recipients for the nine NSF renewable energy initiatives we identified. We generally classified governmental recipients as public and nongovernmental recipients as private.

Figure 87. Percentage of NSF Initiatives Supporting Private or Public Sector Recipients, or Both.

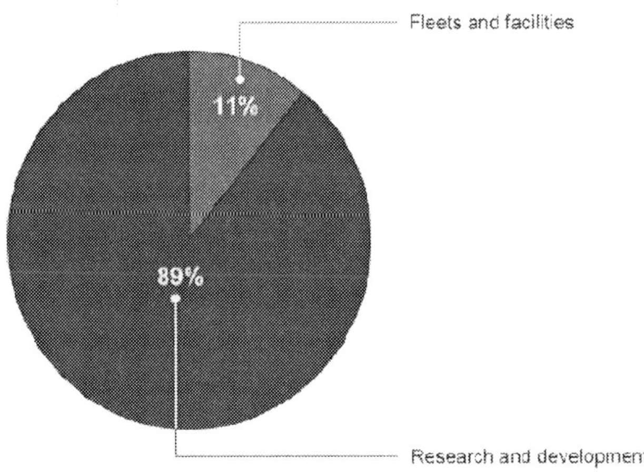

Source: GAO analysis of agency data.

Note: This figure provides data on the federal roles for the nine NSF renewable energy initiatives we identified.

Figure 88. Percentage of NSF Initiatives by Federal Role and Other.

Small Business Administration

The Small Business Administration's (SBA) seven initiatives were fairly evenly distributed in their support for energy sources in our review and most supported multiple energy sources, as figures 89 and 90, respectively, show.

Figure 91 demonstrates that a majority of SBA initiatives supported private sector recipients, while figure 92 shows that more than half supported efforts outside of the four key federal roles in our review.

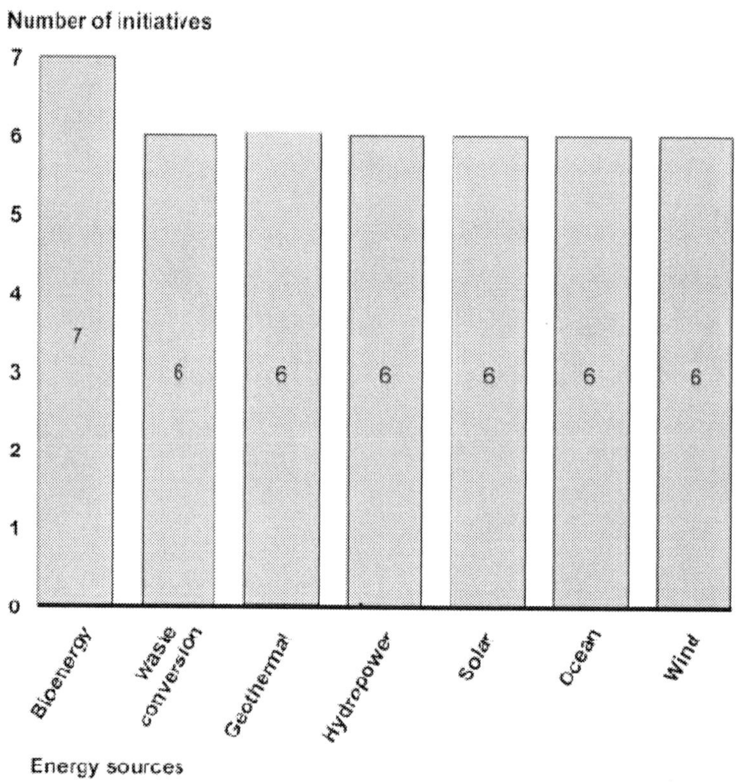

Source: GAO analysis of agency data.

Note: The sum of the initiatives across the renewable energy sources adds to more than the total number of SBA initiatives because initiatives often supported multiple energy sources.

Figure 89. Number of SBA Initiatives Supporting Each Renewable Energy Source.

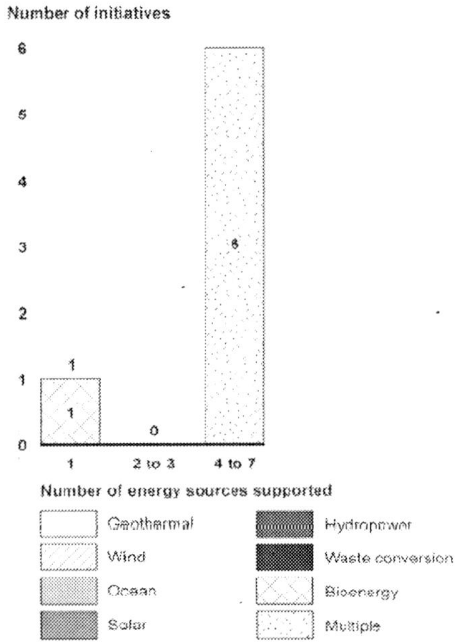

Source: GAO analysis of agency data.

Figure 90. Number of Energy Sources Supported by SBA Initiatives.

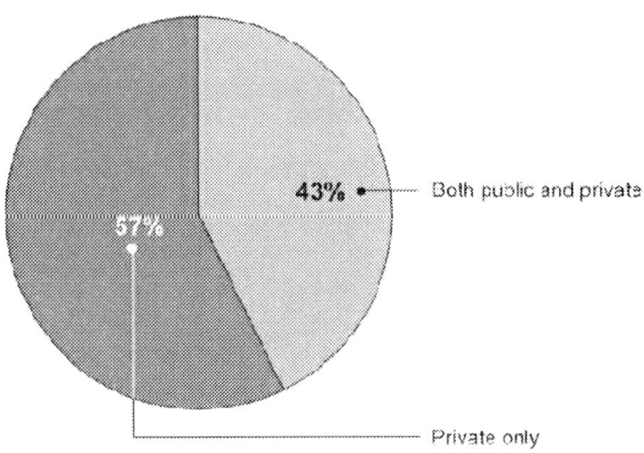

Source: GAO analysis of agency data.

Notes: This figure provides data on the recipients for the seven SBA renewable energy initiatives we identified. We generally classified governmental recipients as public and nongovernmental recipients as private.

Figure 91. Percentage of SBA Initiatives Supporting Private or Public Sector Recipients, or Both.

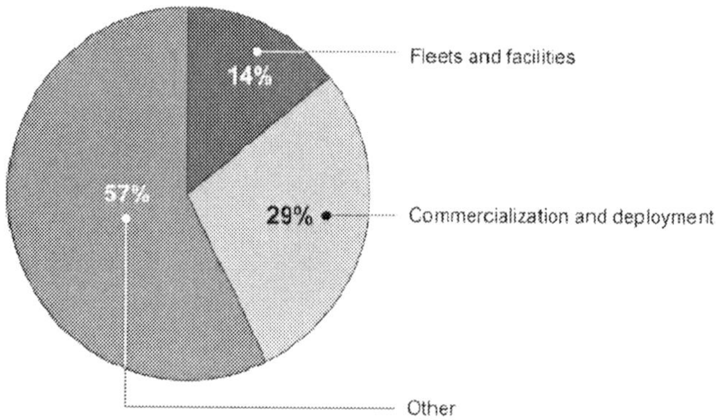

Source: GAO analysis of agency data.

Note: This figure provides data on the federal roles for the seven SBA renewable energy initiatives we identified.

Figure 92. Percentage of SBA Initiatives by Federal Role and Other.

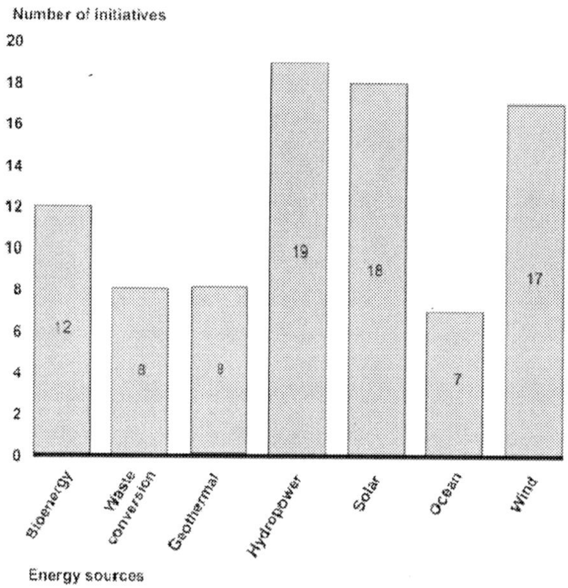

Source: GAO analysis of agency data.

Note: The sum of the initiatives across the renewable energy sources adds to more than the total number of USAID initiatives because initiatives often supported multiple energy sources.

Figure 93. Number of USAID Initiatives Supporting Each Renewable Energy Source.

U.S. Agency for International Development

Figure 93 shows that the U.S. Agency for International Development's (USAID) 25 initiatives most commonly supported hydropower, solar, and wind energy. Also, a majority of USAID's initiatives supported more than one energy source, as shown in figure 94. Figure 95 demonstrates that more than 50 percent of USAID's initiatives supported both public and private sector recipients. Figure 96 shows that a large majority of USAID initiatives supported efforts outside of the four key federal roles in our review, such as training and capacity building.

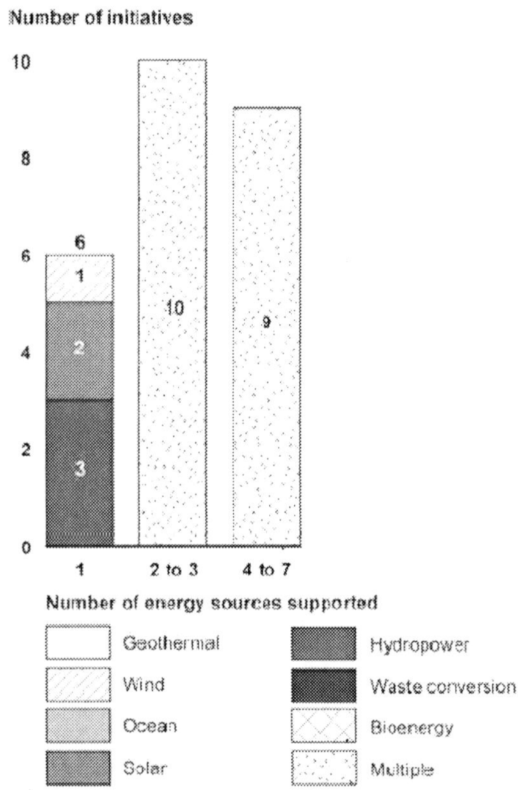

Source: GAO analysis of agency data.

Figure 94. Number of Energy Sources Supported by USAID Initiatives.

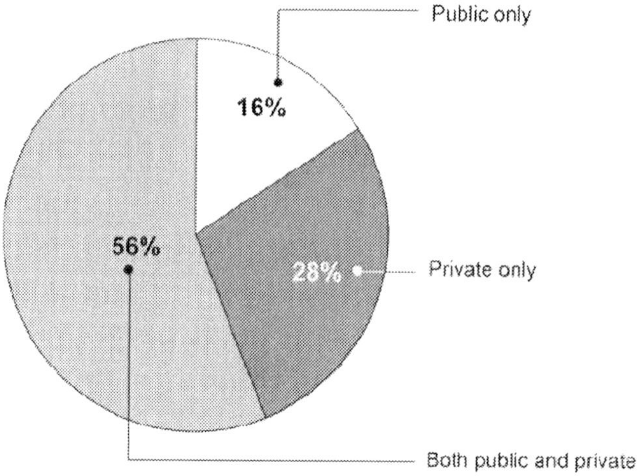

Source: GAO analysis of agency data.

Notes: This figure provides data on the recipients for the 25 USAID renewable energy initiatives we identified. We generally classified governmental recipients as public and nongovernmental recipients as private.

Figure 95. Percentage of USAID Initiatives Supporting Private or Public Sector Recipients, or Both.

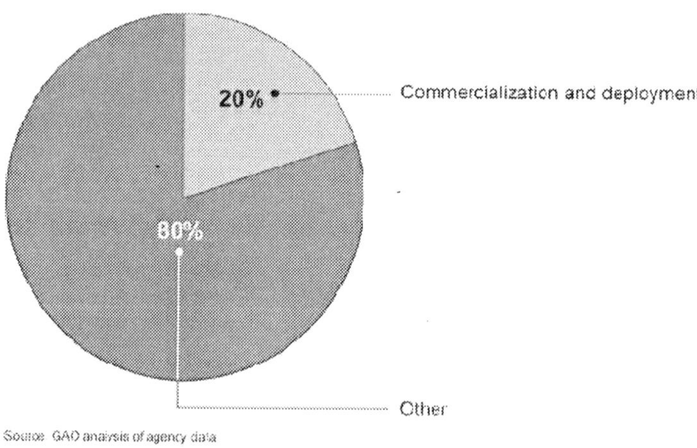

Source: GAO analysis of agency data.

Note: This figure provides data on the federal roles for the 25 USAID renewable energy initiatives we identified.

Figure 96. Percentage of USAID Initiatives by Federal Role and Other.

Agencies' Initiatives across the Four Key Federal Roles, by Energy Source

Table 8 provides information on agencies' initiatives that support bioenergy across the four key federal roles.

Table 8. Number of Agencies' Initiatives That Support Bioenergy across the Four Key Federal Roles

	Research and development		Fleets and facilities		Commercialization and deployment		Regulation, permitting, and compliance	
Agency	Targeted on bioenergy	Supports multiple sources	Targeted on bioenergy	Supports multiple sources	Targeted on bioenergy	Supports multiple sources	Targeted on bioenergy	Supports multiple sources
Commerce	4	2	9	1	0	4	0	1
DOD								
Air Force	2	3	1	2	0	0	0	0
Army	2	5	1	4	0	0	0	1
Marine Corps	1	2	1	1	0	0	0	0
Navy	3	0	1	1	0	0	0	0
Other DOD	4	3	6	3	1	0	0	0
DHS	0	2	4	4	0	0	0	0
DOE	9	10	2	7	0	10	0	0
DOT	8	3	2	1	0	10	1	0
EPA	4	2	1	1	0	4	5	1
FERC	0	0	1	0	0	0	0	0
GSA	0	0	3	1	0	0	0	0
HUD	0	0	0	0	0	0	0	0
Interior	4	1	7	3	2	4	0	2
Justice	0	0	8	2	0	0	0	0
Labor	0	0	1	1	0	0	0	0
NASA	1	0	0	3	0	0	0	0
NSF	0	5	0	0	0	0	0	0
SBA	0	0	1	0	0	2	0	0
State	0	0	1	0	0	3	0	0
Treasury	0	0	3	1	8	12	1	0
USAID	0	0	0	0	0	3	0	0
USDA	20	8	8	4	11	14	0	0
Total	62	46	61	40	22	66	7	5

Source: GAO analysis of agencies' data.

Note: This table does not include data for 22 initiatives for which we identified a federal role, but could not determine the applicable renewable energy sources based on the responses received from the following agencies: the Army (1), DOE (15), HUD (4), the Navy (1), and USDA (1).

As shown in the table, the research and development and fleets and facilities federal roles have the most initiatives targeted on bioenergy specifically, while agencies' commercialization and deployment initiatives that support bioenergy often support other renewable energy sources as well.

Table 9. Number of Agencies' Initiatives That Support Solar Energy across the Four Key Federal Roles

	Research and development		Fleets and facilities		Commercialization and deployment		Regulation, permitting, and compliance	
Agency	Targeted on solar	Supports multiple sources	Targeted on solar	Supports multiple sources	Targeted on solar	Supports multiple sources	Targeted on solar	Supports multiple sources
Commerce	2	4	1	3	0	4	0	1
DOD								
Air Force	2	3	0	6	0	0	0	0
Army	6	3	1	6	0	0	0	1
Marine Corps	1	2	2	4	0	0	0	0
Navy	0	1	0	5	0	0	0	0
Other DOD	8	2	5	6	0	0	0	0
DHS	1	1	3	5	0	0	0	0
DOE	6	11	0	7	2	8	0	0
DOT	0	1	1	2	0	11	1	0
EPA	0	1	0	1	0	1	0	1
FERC	0	2	1	0	0	0	0	1
GSA	0	1	0	4	0	0	0	0
HUD	0	0	0	0	1	8	0	0
Interior	0	3	2	10	1	7	1	13
Justice	0	0	0	3	0	0	0	0
Labor	0	0	0	1	0	0	0	0
NASA	2	0	1	4	0	0	0	0
NSF	1	7	0	0	0	0	0	0
SBA	0	0	0	0	0	2	0	0
State	0	0	0	3	0	3	0	0
Treasury	0	0	0	0	1	13	0	0
USAID	0	0	0	0	0	4	0	0
USDA	0	4	2	8	0	9	0	1
Total	29	46	19	78	5	70	2	18

Source: GAO analysis of agencies' data.

Note: This table does not include data for 22 initiatives for which we identified a federal role, but could not determine the applicable renewable energy sources based on the responses received from the following agencies: the Army (1), DOE (15), HUD (4), the Navy (1), and USDA (1).

Table 9 provides information on agencies' initiatives that support solar energy across the four key federal roles. As shown in the table, the research and development federal role has the most initiatives targeted on solar energy specifically. Agencies' fleets and facilities and commercialization and deployment initiatives generally support solar energy along with other energy sources.

Table 10. Number of Agencies' Initiatives That Support Wind Energy across the Four Key Federal Roles

	Research and development		Fleets and facilities		Commercialization and deployment		Regulation, permitting, and compliance	
Agency	Targeted on wind	Supports multiple sources	Targeted on wind	Supports multiple sources	Targeted on wind	Supports multiple sources	Targeted on wind	Supports multiple sources
Commerce	1	4	2	3	0	5	0	1
DOD								
Air Force	0	2	0	3	0	0	0	0
Army	0	3	0	4	0	0	0	1
Marine Corps	0	2	0	1	0	0	0	0
Navy	0	0	0	3	0	0	0	0
Other DOD	0	0	0	5	0	0	0	0
DHS	0	1	1	6	0	0	0	1
DOE	2	9	0	8	1	8	0	0
DOT	0	1	1	3	0	11	0	0
EPA	0	1	0	1	0	1	0	1
FERC	0	2	0	0	0	0	0	1
GSA	0	0	0	4	0	0	0	0
HUD	0	0	0	0	0	8	0	0
Interior	1	4	1	10	0	6	3	14
Justice	0	0	1	4	0	0	0	0
Labor	0	0	1	1	0	0	0	0
NASA	0	0	0	3	0	0	0	0
NSF	0	4	1	0	0	0	0	0
SBA	0	0	0	0	0	2	0	0
State	0	0	0	2	0	3	0	0
Treasury	0	0	2	1	0	12	0	0
USAID	0	0	0	0	0	3	0	0
USDA	1	4	0	8	0	9	0	1
Total	5	37	10	70	1	68	3	20

Source: GAO analysis of agencies' data.

Note: This table does not include data for 22 initiatives for which we identified a federal role, but could not determine the applicable renewable energy sources based on the responses received from the following agencies: the Army (1), DOE (15), HUD (4), the Navy (1), and USDA (1).

Table 10 provides information on agencies' initiatives that support wind energy across the four key federal roles. As shown in the table, agencies' initiatives generally do not target wind energy specifically. Rather, agencies' initiatives tend to support wind energy along with other energy sources across the four key federal roles.

Table 11. Number of Agencies' Initiatives That Support Geothermal Energy across the Four Key Federal Roles

	Research and development		Fleets and facilities		Commercialization and deployment		Regulation, permitting, and compliance	
Agency	Targeted on geothermal	Supports multiple sources	Targeted on geothermal	Supports multiple sources	Targeted on geothermal	Supports multiple sources	Targeted on geothermal	Supports multiple sources
Commerce	0	1	0	2	0	4	0	1
DOD								
Air Force	0	0	0	5	0	0	0	0
Army	0	2	0	5	0	0	0	1
Marine Corps	0	2	0	3	0	0	0	0
Navy	0	0	1	4	0	0	0	0
Other DOD	0	0	0	6	0	0	0	0
DHS	0	0	0	1	0	0	0	0
DOE	4	5	0	6	0	7	0	0
DOT	0	0	0	2	0	11	0	0
EPA	0	0	0	1	0	1	0	1
FERC	0	0	0	0	0	0	0	0
GSA	1	1	0	3	0	0	0	0
HUD	0	0	0	0	0	6	0	0
Interior	1	0	0	6	1	6	1	13
Justice	0	0	0	1	0	0	0	0
Labor	0	0	0	0	0	0	0	0
NASA	0	0	0	1	0	0	0	0
NSF	0	1	0	0	0	0	0	0
SBA	0	0	0	0	0	2	0	0
State	0	0	0	2	0	2	0	0
Treasury	0	0	0	0	0	12	0	0
USAID	0	0	0	0	0	2	0	0
USDA	0	0	0	4	0	8	1	0
Total	6	12	1	52	1	61	2	16

Source: GAO analysis of agencies' data.

Note: This table does not include data for 22 initiatives for which we identified a federal role, but could not determine the applicable renewable energy sources based on the responses received from the following agencies: the Army (1), DOE (15), HUD (4), the Navy (1), and USDA (1).

Table 11 provides information on agencies' initiatives that support geothermal energy across the four key federal roles. As shown in the table, most agencies' geothermal energy initiatives are fleets and facilities or commercialization and deployment initiatives that support geothermal energy along with other energy sources.

Table 12. Number of Agencies' Initiatives That Support Waste Conversion across the Four Key Federal Roles

Agency	Research and development		Fleets and facilities		Commercialization and deployment		Regulation, permitting, and compliance	
	Targeted on waste conversion	Supports multiple sources	Targeted on waste conversion	Supports multiple sources	Targeted on waste conversion	Supports multiple sources	Targeted on waste conversion	Supports multiple sources
Commerce	0	1	0	1	1	4	0	1
DOD								
Air Force	0	2	0	3	0	0	0	0
Army	1	5	0	4	0	0	0	1
Marine Corps	0	2	0	1	0	0	0	0
Navy	0	1	0	1	0	0	0	0
Other DOD	0	2	0	2	0	0	0	0
DHS	0	1	0	2	0	0	0	0
DOE	0	7	0	3	0	6	0	0
DOT	0	3	0	0	0	0	0	0
EPA	0	2	0	1	1	3	0	1
FERC	0	0	0	0	0	0	0	0
GSA	0	0	0	1	0	0	0	0
HUD	0	0	0	0	0	0	0	0
Interior	0	0	0	2	1	4	0	1
Justice	0	0	0	1	0	0	0	0
Labor	0	0	0	0	0	0	0	0
NASA	0	0	0	1	0	0	0	0
NSF	0	3	0	0	0	0	0	0
SBA	0	0	0	0	0	2	0	0
State	0	0	0	0	0	3	0	0
Treasury	0	0	0	0	0	12	0	0
USAID	0	0	0	0	0	1	0	0
USDA	1	6	0	2	2	13	0	0
Total	**2**	**35**	**0**	**25**	**5**	**48**	**0**	**4**

Source: GAO analysis of agencies' data.

Note: This table does not include data for 22 initiatives for which we identified a federal role, but could not determine the applicable renewable energy sources based on the responses received from the following agencies: the Army (1), DOE (15), HUD (4), the Navy (1), and USDA (1).

Table 13. Number of Agencies' Initiatives That Support Hydropower across the Four Key Federal Roles

	Research and development		Fleets and facilities		Commercialization and deployment		Regulation, permitting, and compliance	
Agency	Targeted on hydropower	Supports multiple sources	Targeted on hydropower	Supports multiple sources	Targeted on hydropower	Supports multiple sources	Targeted on hydropower	Supports multiple sources
Commerce	0	3	0	1	0	4	1	1
DOD								
Air Force	0	0	0	0	0	0	0	0
Army	0	1	0	2	0	0	0	1
Marine Corps	0	1	0	0	0	0	0	0
Navy	0	0	0	0	0	0	0	0
Other DOD	0	0	1	0	0	0	0	0
DHS	0	0	0	1	0	0	0	0
DOE	1	6	0	4	1	8	0	0
DOT	0	0	0	1	0	0	0	0
EPA	0	1	0	0	0	1	0	1
FERC	0	0	0	0	0	0	1	1
GSA	0	0	0	0	0	0	0	0
HUD	0	0	0	0	0	0	0	0
Interior	3	1	0	4	2	2	3	5
Justice	0	0	0	0	0	0	0	0
Labor	0	0	0	0	0	0	0	0
NASA	0	0	0	0	0	0	0	0
NSF	0	2	0	0	0	0	0	0
SBA	0	0	0	0	0	2	0	0
State	0	0	0	0	0	2	0	0
Treasury	0	0	0	0	0	11	0	0
USAID	0	0	0	0	1	3	0	0
USDA	0	0	0	3	0	7	0	1
Total	4	15	1	16	4	40	5	10

Source: GAO analysis of agencies' data.

Notes: This table does not include data for 22 initiatives for which we identified a federal role, but could not determine the applicable renewable energy sources based on the responses received from the following agencies: the Army (1), DOE (15), HUD (4), the Navy (1), and USDA (1). In addition, some agencies implemented initiatives that provided important support for hydropower, such the energy generated by the U.S. Army Corps of Engineers' and Bureau of Reclamation's hydroelectric facilities. These activities were classified in our "other" category and are not included in the table, which focuses on the four key federal roles.

Table 12 provides information on agencies' initiatives that support waste conversion across the four key federal roles. As shown in the table, the commercialization and deployment role has the most initiatives supporting waste conversion, and agencies' initiatives tend to support waste conversion along with other energy sources across the four key federal roles.

Table 13 provides information on agencies' initiatives that support hydropower across the four key federal roles. As shown in the table, the commercialization and deployment role has the most initiatives supporting hydropower, and the majority of hydropower initiatives across the four roles support other sources as well.

Table 14. Number of Agencies' Initiatives That Support Ocean Energy across the Four Key Federal Roles

	Research and development		Fleets and facilities		Commercialization and deployment		Regulation, permitting, and compliance	
Agency	Targeted on ocean	Supports multiple sources	Targeted on ocean	Supports multiple sources	Targeted on ocean	Supports multiple sources	Targeted on ocean	Supports multiple sources
Commerce	0	2	0	1	0	5	1	1
DOD								
Air Force	0	1	0	0	0	0	0	0
Army	0	1	0	0	0	0	0	1
Marine Corps	0	1	0	0	0	0	0	0
Navy	1	1	0	1	0	0	0	0
Other DOD	4	0	0	1	0	0	0	0
DHS	0	2	0	0	0	0	0	1
DOE	1	5	0	3	0	3	0	0
DOT	0	0	0	1	0	0	0	0
EPA	0	0	0	0	0	0	0	0
FERC	0	0	0	0	0	0	0	1
GSA	0	0	0	0	0	0	0	0
HUD	0	0	0	0	0	0	0	0
Interior	0	2	0	0	0	3	0	12
Justice	0	0	0	0	0	0	0	0
Labor	0	0	0	0	0	0	0	0
NASA	0	0	0	0	0	0	0	0
NSF	0	2	0	0	0	0	0	0
SBA	0	0	0	0	0	2	0	0
State	0	0	0	0	0	1	0	0
Treasury	0	0	0	0	0	9	0	0
USAID	0	0	0	0	0	1	0	0
USDA	0	0	0	1	0	3	0	0
Total	6	17	0	8	0	27	1	16

Source: GAO analysis of agencies' data.

Note: This table does not include data for 22 initiatives for which we identified a federal role, but could not determine the applicable renewable energy sources based on the responses received from the following agencies: the Army (1), DOE (15), HUD (4), the Navy (1), and USDA (1).

Table 14 provides information on agencies' initiatives that support ocean energy across the four key federal roles. As shown in the table, the commercialization and deployment role has the most initiatives supporting ocean energy, and the majority of ocean energy initiatives across the four roles support other sources as well.

APPENDIX III: AGENCY INFORMATION ON THE EXTENT TO WHICH THE AMERICAN RECOVERY AND REINVESTMENT ACT OF 2009 ESTABLISHED, EXPANDED, OR MODIFIED INITIATIVES

This appendix provides information on the extent to which the American Recovery and Reinvestment Act of 2009 (Recovery Act) established, expanded, or modified renewable energy initiatives. Specifically, agencies reported that the Recovery Act established, expanded, or modified 157 initiatives—nearly one-quarter of all renewable energy initiatives governmentwide.[1] These effects fall into four groupings:

- *Initiatives that were established by the Recovery Act.* Across all of the agencies in our review, the Recovery Act established 20 initiatives. These initiatives mostly supported the commercialization and deployment federal role. For example, the Department of the Treasury's Payments for Specific Energy Property in Lieu of Tax Credits were established under Section 1603 of the Recovery Act to provide cash grants to eligible applicants who place specified energy property—primarily renewable energy systems—in service for use in a trade or business.[2]
- *Initiatives that received additional funding under the Recovery Act.* There were 106 initiatives that existed prior to the Recovery Act that received additional funding for renewable energy activities. Receiving additional funding was the most commonly experienced type of Recovery Act effect we identified among the initiatives. Approximately one-third of these initiatives (36) were implemented by the Department of Energy, primarily involving research and development of new renewable energy technologies. A number of Department of Defense fleets and facilities efforts also received additional funding under the Recovery Act. In addition, the General

Services Administration's efforts to purchase vehicles for federal agency fleets included a vehicle replacement program, which received Recovery Act funds for the purchase of alternative fuel vehicles. Under this program, federal agencies could exchange existing vehicles for more fuel efficient models at no cost to them.

- *Initiatives that were extended or whose scope was changed by the Recovery Act.* There were 26 initiatives that experienced an extension or change in their scope of authority under the Recovery Act. More than 80 percent of these initiatives (21) were implemented by the departments of Energy or the Treasury. For example, for certain Department of the Treasury initiatives, the Recovery Act increased the bond authority available for allocation by the Internal Revenue Service to support renewable energy activities.
- *Initiatives that were indirectly affected by the Recovery Act.* There were 19 initiatives that experienced indirect effects from the Recovery Act aside from those mentioned above. For example, for some initiatives, agencies reported that staff workloads increased as a result of the Recovery Act even though the initiatives did not receive direct funding from the Recovery Act.

Table 15 shows the initiatives that the Recovery Act affected at each agency and the type of effects they experienced based on our analysis of agency-provided data.

Table 15. Initiatives Established, Expanded, or Modified by the Recovery Act, by Agency

Agency (number of initiatives)	Initiative name	Initiative established by Recovery Act	Initiative received additional funding from Recovery Act	Initiative extended or scope changed by Recovery Act	Indirect or other impacts
Department of Agriculture (11)	Alternative Fuels and Fleet Efficiency Program		X		
	Recovery Act Wood to Energy Biomass Activity	X			

Table 15. (Continued)

Agency (number of initiatives)	Initiative name	Initiative established by Recovery Act	Initiative received additional funding from Recovery Act	Initiative extended or scope changed by Recovery Act	Indirect or other impacts
	Agricultural Research Service Alternative Fuel Vehicle Acquisition and Use		X		
	Biomass Crop Assistance Program[a]			X	
	Business and Industry Guaranteed Loan Program		X		
	Energy Management Program		X		
	Farm to Fly: Sustainable Aviation Biofuel		X		
	Forest Service Alternative Fuel Vehicle Acquisition and Use		X		
	Forest Service-Funded Renewable Power Generation		X		
	Natural Resources Conservation Service Alternative Fuel Vehicle Acquisition and Use		X		
	Woody Biomass Utilization Grant Program		X		
Department of Commerce (5)	Environmentally-Sustainable Development Investment Priority		X		
	Licensing Authority for Ocean Thermal Energy Conversion Facilities				X

Agency (number of initiatives)	Initiative name	Initiative established by Recovery Act	Initiative received additional funding from Recovery Act	Initiative extended or scope changed by Recovery Act	Indirect or other impacts
	Measurement Science for Net Zero Energy, High-Performance Buildings		X		
	National Institute of Standards and Technology Smart Grid Research		X		
	National Institute of Standards and Technology-Funded Renewable Power Generation		X		
Department of Defense, Air Force (2)	Air Force Non-tactical Alternative Fuel Vehicle Acquisition and Use		X		
	Air Force Renewable Energy Certificate and/or Direct Renewable Power Purchases		X		
Department of Defense, Army (5)	Army Non-tactical Alternative Fuel Vehicle Acquisition and Use		X		
	Army Tactical Equipment Procurement and Deployment		X		
	Army-Funded Small Scale Renewable Power Generation		X		
	Corps Hydropower Program		X		
	Permitting Efficiencies for Regulatory Programs				X

Table 15. (Continued)

Agency (number of initiatives)	Initiative name	Initiative established by Recovery Act	Initiative received additional funding from Recovery Act	Initiative extended or scope changed by Recovery Act	Indirect or other impacts
Department of Defense, Marine Corps (2)	Marine Corps-Funded Medium Scale Renewable Power Generation		X		
	Marine Corps-Funded Small Scale Renewable Power Generation		X		
Department of Defense, Navy (5)	Alternative Fuels Program		X		
	Navy Non-tactical Alternative Fuel Vehicle Acquisition and Use		X		
	Navy-Funded Medium Scale Renewable Power Generation		X		
	Navy-Funded Small Scale Renewable Power Generation		X		
	Other Power Generation Research		X		
Department of Defense, other Department of Defense	Defense Logistics Agency Alternative Fuel Development and Testing		X		
	Defense Logistics Agency Portable Power Generation Research		X		
components (8)	Defense Logistics Agency Renewable Energy Outreach				X
	Defense Wide Manufacturing Science and Technology Program		X		

Agency (number of initiatives)	Initiative name	Initiative established by Recovery Act	Initiative received additional funding from Recovery Act	Initiative extended or scope changed by Recovery Act	Indirect or other impacts
	Energy Conservation Investment Program		X		
	Environmental Security Technology Certification Program – Energy and Water Program		X		
	National Security Agency-Funded Small Scale Renewable Power Generation		X		
	Net Zero Plus Joint Capability Technology Demonstration Program		X		
Department of Energy (49)	Agile Delivery of Electrical Power Technology	X			
	Advanced Research Projects Agency-Energy Funding Opportunity Announcement 1	X			
	Bioenergy Research Centers		X	X	
	Biomass and Biorefinery Systems Infrastructure		X	X	
	Bonneville Power Administration Direct Funding			X	
	Bonneville Power Administration Fish and Wildlife			X	
	Bonneville Power Administration Wind Integration			X	
	Clean Energy Transmission and Reliability		X		
	Concentrated Solar Power		X		

Table 15. (Continued)

Agency (number of initiatives)	Initiative name	Initiative established by Recovery Act	Initiative received additional funding from Recovery Act	Initiative extended or scope changed by Recovery Act	Indirect or other impacts
	Conventional Hydropower Technology Development		X		
	Conversion Technologies		X	X	
	Electrofuels	X			
	Emerging Technologies		X		
	Energy Efficiency and Conservation Block Grants	X			
	Energy Frontier Research Centers		X		
	Enhanced Geothermal Systems		X		
	Feedstocks		X	X	
	Federal Energy Management Program -Planning, Reporting and Evaluation		X		
	Federal Energy Management Program -Project Financing				X
	Federal Energy Management Program -Technical Guidance and Assistance Fuel Cell Systems Research & Development		X		
	Funding the work of the Western Governors Association to identify Western Renewable Energy Zone identification		X		
	Grid-Scale Rampable Intermittent Dispatchable Storage		X	X	

Agency (number of initiatives)	Initiative name	Initiative established by Recovery Act	Initiative received additional funding from Recovery Act	Initiative extended or scope changed by Recovery Act	Indirect or other impacts
	Innovative Exploration Technologies	X			
	Integrated Biorefineries	X			
	Low Temperature and Co-Produced Geothermal Technologies Resources		X	X	
	Office of Electricity Delivery and Energy Reliability Energy Storage		X		
	Office of Electricity Delivery and Energy Reliability Permitting, Siting, and Analysis		X		
	Photovoltaic Research & Development		X		
	Power Marketing Administrations—Transmission		X		
	Small Business Innovation Research/Small Business Technology Transfer - Advanced Solar Technology Topic Area			X	
	Small Business Innovation Research/Small Business Technology Transfer - Catalysis Topic Area		X		

Table 15. (Continued)

Agency (number of initiatives)	Initiative name	Initiative established by Recovery Act	Initiative received additional funding from Recovery Act	Initiative extended or scope changed by Recovery Act	Indirect or other impacts
	Small Business Innovation Research/Small Business Technology Transfer - Geothermal Energy Technology Development Topic Area		X		
	Small Business Innovation Research/Small Business Technology Transfer - Novel Membrane and Electrode Development for Advanced Electrochemical Energy Storage Topic Area		X		
	Small Business Innovation Research/Small Business Technology Transfer - Production of Biofuels from Cellulosic Biomass Topic Area		X		
	Small Business Innovation Research/Small Business Technology Transfer - Technologies for Clean Fuels and Hydrogen from Coal Topic Area		X		

Agency (number of initiatives)	Initiative name	Initiative established by Recovery Act	Initiative received additional funding from Recovery Act	Initiative extended or scope changed by Recovery Act	Indirect or other impacts
	Small Business Innovation Research/Small Business Technology Transfer - Technologies Related to Energy Storage for Hybrid Plug-in Electric Vehicles Topic Area		X		
	Small Business Innovation Research/Small Business Technology Transfer - Wind Energy Technology Development Topic Area		X		
	Smart Grid Research and Development		X		
	Solar Decathlon		X		X
	State Energy Program		X		
	Technology Advancement and Outreach Activities				X
	Title XVII Section 1705 Loan Program	X			
	Tribal Energy Program		X	X	
	Vehicle Technology Deployment		X		
	Weatherization Assistance Program		X		
	Wind Energy - Offshore Wind		X		
	Wind Energy - Technology Application		X		
	Wind Energy - Technology Viability		X		

Table 15. (Continued)

Agency (number of initiatives)	Initiative name	Initiative established by Recovery Act	Initiative received additional funding from Recovery Act	Initiative extended or scope changed by Recovery Act	Indirect or other impacts
Department of Homeland Security (7)	Customs and Border Protection Alternative Fuel Vehicle Acquisition and Use		X		
	Customs and Border Protection-Funded Renewable Power Generation		X		
	Immigration and Customs Enforcement Alternative Fuel Vehicle Acquisition and Use		X		
	Coast Guard Alternative Fuel Vehicle Acquisition and Use		X		
	Coast Guard Third Party-Financed Medium Scale Renewable Power Generation				X
	Coast Guard-Funded Medium Scale Renewable Power Generation		X		X
	Coast Guard-Funded Small Scale Renewable Power Generation		X		X
Department of Housing and Urban Development (3)	Capital Fund Recovery Act Competitive Grant Program	X			
	Green Retrofit Program for Multifamily Housing	X			
	Neighborhood Stabilization Program 2	X			

Agency (number of initiatives)	Initiative name	Initiative established by Recovery Act	Initiative received additional funding from Recovery Act	Initiative extended or scope changed by Recovery Act	Indirect or other impacts
Department of the Interior (23)	Bureau of Land Management-Funded Renewable Power Generation		X		
	Conservation Planning Assistance Program				X
	Consultation and Habitat Conservation Planning				X
	Cooperative Water Program		X		
	External Renewable Energy Program				X
	External Renewable Energy Training		X	X	
	Fish and Wildlife Service Alternative Fuel Vehicle Acquisition and Use		X		
	Fish and Wildlife Service-Funded Renewable Power Generation		X		
	Hazardous Fuels Reduction Program		X		
	Housing Improvement Program: Recovery Act Renewable Energy Projects	X			
	Indian Affairs-Funded Renewable Power Generation		X		
	Interior Renewable Energy Certificate and/or Direct Renewable Power Purchases				X
	Interior Renewable Energy Policy Development and Management				X
	Klamath Dam Removal Studies		X		

Table 15. (Continued)

Agency (number of initiatives)	Initiative name	Initiative established by Recovery Act	Initiative received additional funding from Recovery Act	Initiative extended or scope changed by Recovery Act	Indirect or other impacts
	National Park Service Alternative Fuel Vehicle Acquisition and Use		X		
	National Park Service-Funded Renewable Power Generation		X		
	Recovery Act Renewable Energy Efforts		X		
	Solar Energy Authorizations and Operations on Bureau of Land Management Public Lands				X
	Solar Energy Programmatic Environmental Impact Statement	X			
	U.S. Geological Survey Alternative Fuel Vehicle Acquisition and Use		X		
	U.S. Geological Survey Third Party-Financed Renewable Power Generation		X		
	U.S. Geological Survey-Funded Renewable Power Generation		X		
	WaterSMART Grant program (formerly known as Challenge Grants)		X	X	
Department of Justice (5)	Bureau of Alcohol, Tobacco, Firearms and Explosives Alternative Fuel Vehicle Acquisition and Use		X		

Agency (number of initiatives)	Initiative name	Initiative established by Recovery Act	Initiative received additional funding from Recovery Act	Initiative extended or scope changed by Recovery Act	Indirect or other impacts
	Drug Enforcement Administration Alternative Fuel Vehicle Acquisition and Use		X		
	Federal Bureau of Investigation Alternative Fuel Vehicle Acquisition and Use		X		
	Justice Management Division Alternative Fuel Vehicle Acquisition and Use		X		
	U.S. Marshals Service Alternative Fuel Vehicle Acquisition and Use		X		
Department of Labor (4)	Energy Training Partnership Grant Program	X			
	Green Capacity Building Grants	X			
	State Energy Sector Partnership and Training Grants	X			
	State Labor Market Information Improvement Grants		X		
Department of Transportation (6)	Federal Aviation Administration Alternative Fuel Vehicle Acquisition and Use		X		
	Fixed Guideway Modernization		X		
	Formula Grants for Other than Urbanized Areas		X		
	Major Capital Investments (New Starts & Small Starts)		X		
	Transit Investments for Greenhouse Gas and Energy Reduction	X			

Table 15. (Continued)

Agency (number of initiatives)	Initiative name	Initiative established by Recovery Act	Initiative received additional funding from Recovery Act	Initiative extended or scope changed by Recovery Act	Indirect or other impacts
	Urbanized Area Formula Program		X		
Department of the Treasury (12)	Accelerated Depreciation Recovery Periods for Specific Energy Property			X	
	Credit for Clean Fuel Burning Vehicles			X	
	Credit for Clean-Fuel Burning Vehicle Refueling Property			X	
	Credit for Holding New Clean Renewable Energy Bonds			X	
	Credit for Holding Qualified Energy Conservation Bonds			X	
	Credit for Residential Energy Efficient Property			X	
	Direct Payment in Lieu of a Credit for Holding New Clean Renewable Energy Bonds			X	
	Direct Payment in Lieu of a Credit for Holding Qualified Energy Conservation Bonds			X	
	Energy Investment Credit			X	
	Energy Production Credit			X	
	Payments for Specific Energy Property in Lieu of Tax Credits	X			
	Qualifying Advanced Energy Project Credit	X			

Agency (number of initiatives)	Initiative name	Initiative established by Recovery Act	Initiative received additional funding from Recovery Act	Initiative extended or scope changed by Recovery Act	Indirect or other impacts
General Services Administration (5)	Geothermal Space Conditioning Technology Acceleration Program				X
	Green Proving Ground Program				X
	General Services Administration Fleet Purchasing		X		
	General Services Administration Renewable Power Generation Data Collection and Reporting				X
	General Services Administration-Funded Renewable Power Generation		X		
National Aeronautics and Space Administration (1)	National Aeronautics and Space Administration Alternative Fuel Vehicle Acquisition and Use		X		
Small Business Administration (4)	Certified Development Company /Section 504 Loans		X	X	
	Energy Grants for Green Technology				X
	Energy Savings Debenture			X	
	Entrepreneurial Mentor Corps Pilot Program	X			

Source: GAO analysis of agencies' data.

a The Recovery Act provided the Department of Agriculture with the authority to use Biomass Crop Assistance Program funds for technical assistance. However, in commenting on a draft of this report, the Department of Agriculture indicated that, by the time it needed to use program funds for technical assistance, this authority had expired and no program funds were used to provide technical assistance.

Appendix IV: Agency Information on Expiring Initiatives

This appendix provides information reported by agencies on initiatives that will or have expired, in full or in part, due to an expiration of legislative authority, depletion of available appropriations, or some other expiration under the law as written as of fall of 2011. This appendix includes information on initiatives that expired before the start of or during fiscal year 2010, or will expire after fiscal year 2010. However, all of these initiatives met our criteria as fiscal year 2010 renewable energy initiatives in some way and are included in the 679 federal renewable energy initiatives for which we provide data in this report. For example, although the Conservation Security Program was not re-authorized in the Food, Conservation, and Energy Act of 2008 and is no longer available, contracts written while the program was still authorized can span up to 10 years, and some contracts were still being implemented in fiscal year 2010. Table 16 provides information on these initiatives, including the agencies implementing them and information reported by the agencies on their expiration. For additional information on the following initiatives, including their descriptions and associated renewable energy sources, federal roles, and recipients, see GAO-12-259SP.

Table 16. Expiring Initiatives

Initiative (agency)	Expiration information
Department of Agriculture	
Recovery Act Wood to Energy Biomass Activity (Forest Service)	This initiative was established by and ended with the AmericanRecovery and Reinvestment Act of 2009.
Conservation Security Program (Natural Resources Conservation Service)	This program no longer exists.
Small Business Innovation Research Program: Aquaculture Topic Area (National Institute of Food and Agriculture)	Authorization for these programs is contingent upon reauthorization of the Small Business Innovation Development Act of 1982, as amended (15 U.S.C. 638).
Small Business Innovation Research Program: Biofuels and Biobased Products Topic Area (National Institute of Food and Agriculture)	
Small Business Innovation Research Program: Forests and Related Resources Topic Area (National Institute of Food and Agriculture)	

Initiative (agency)	Expiration information
Small Business Innovation Research Program: Plant Production and Protection-Biology Topic Area (National Institute of Food and Agriculture)	
Small Business Innovation Research Program: Plant Production and Protection-Engineering Topic Area (National Institute of Food and Agriculture)	
Small Business Innovation Research Program: Rural Development Topic Area (National Institute of Food and Agriculture)	
Small Business Innovation Research Program: Small and Mid-Size Farms Topic Area (National Institute of Food and Agriculture)	
Community Wood to Energy Program (Forest Service)	Authorization for these programs expires with the end of the Food, Conservation, and Energy Act of 2008. However, the Department of Agriculture reported that should funding remain available after the expiration of the act, it has the authority to continue program operations based upon existing regulations.
Forest Biomass for Energy Competitive Research and Development Program (Forest Service)	
Woody Biomass Utilization Grant Program (Forest Service)	
Biomass Crop Assistance Program (Farm Service Agency)	
Feedstock Flexibility Program for Bioenergy Producers (Farm Service Agency)	
New Era Rural Technology Program (National Institute of Food and Agriculture)	
Sun Grant Program (National Institute of Food and Agriculture)	
Biodiesel Fuel Education Program (National Institute of Food and Agriculture and Office of the Chief Economist)	
Biomass Research and Development Initiative (National Institute of Food and Agriculture)	
Conservation Innovation Grant Program (Natural Resources Conservation Service)	
Environmental Quality Incentives Program (Natural Resources Conservation Service)	
Advanced Biofuel Payment Program (Rural Business Service)	

Table 16. (Continued)

Initiative (agency)	Expiration information
Biorefinery Assistance Program (Rural Business Service)	
Repowering Assistance Program (Rural Business Service)	
Rural Energy for America Program (formerly the Renewable Energy Systems and Energy Efficiency Improvements Program) (Rural Business Service)	
Digester Project (Office of the Secretary)	These initiatives are implemented through the other programs inthis table that will expire at the end of the Food, Conservation, and Energy Act of 2008.
USDA / Navy Memorandum of Understanding Project (Office of the Secretary)	
USDA Energy Coordination and Collaboration (Office of the Secretary)	
Farm to Fly: Sustainable Aviation Biofuel (Rural Business Service)	
Department of Commerce	
Global Climate Change Mitigation Incentive Fund (Economic Development Administration)	The Global Climate Change Mitigation Incentive Fund is not a standing program, but rather a congressionally-directed funding stream, which requires annual submission of a report/spending plan to Congress summarizing how the funds will be used.
Department of Defense	
Net Zero Plus Joint Capability Technology Demonstration Program (Office of the Secretary of Defense)	This was a one-time demonstration program with no further appropriations.
Defense Production Act Title III Program (Office of the Secretary of Defense)	Authority to execute Defense Production Act Title III projects expires September 30, 2014, unless the act is amended.
Department of Energy	
Advanced Research Projects Agency-Energy Funding Opportunity Announcement 1 (Advanced Research Projects Agency-Energy)	The Advanced Research Projects Agency-Energy's authorizing legislation, the America Creating Opportunities to Meaningfully Promote Excellence in Technology, Education, and Science Reauthorization Act of 2010, was recently extended to until 2013.
Agile Delivery of Electrical Power Technology (Advanced Research Projects Agency-Energy)	
Electrofuels (Advanced Research Projects Agency-Energy)	

Initiative (agency)	Expiration information
Grid-Scale Rampable Intermittent Dispatchable Storage (Advanced Research Projects Agency-Energy)	
Hydrogen Fuel R&D (Office of Energy Efficiency and Renewable Energy)	The Energy Policy Act of 2005 and the Energy Independence and Security Act of 2007 provide authorization for these activities through fiscal year 2015 or 2020 depending upon the activity.
Hydrogen and Fuel Cell Technologies Crosscutting Activities (Office of Energy Efficiency and Renewable Energy)	
Fuel Cell Systems R&D (Office of Energy Efficiency and Renewable Energy)	
Energy Efficiency and Conservation Block Grants (Office of Energy Efficiency and Renewable Energy)	This program is currently funded through American Recovery and Reinvestment Act of 2009 funding and not through annual appropriations. Without inclusion in future annual appropriations, the program terminates on December 31, 2011.[b]
Title XVII Section 1705 Loan Program (Loan Programs Office)	The American Recovery and Reinvestment Act of 2009 added Section 1705 to the Energy Policy Act of 2005. There is a sunset date of September 30, 2011 on this initiative.[b]
Small Business Innovation Research/Small Business Technology Transfer - Wind Energy Technology Development Topic Area (multiple)	Continuation of the Small Business Innovation Research/Small Business Technology Transfer programs is contingent upon their reauthorization.
Small Business Innovation Research/Small Business Technology Transfer - Technologies Related to Energy Storage for Hybrid Plug-in Electric Vehicles Topic Area (multiple)	
Small Business Innovation Research/Small Business Technology Transfer - Technologies for Clean Fuels and Hydrogen from Coal (multiple)	
Small Business Innovation Research/Small Business Technology Transfer - Production of Biofuels from Cellulosic Biomass Topic Area (multiple)	
Small Business Innovation Research/Small Business Technology Transfer - Novel Membrane and Electrode Development for Advanced Electrochemical Energy Storage Topic Area (multiple)	

Table 16. (Continued)

Initiative (agency)	Expiration information
Small Business Innovation Research/Small Business Technology Transfer - Geothermal Energy Technology Development Topic Area (multiple)	
Small Business Innovation Research/Small Business Technology Transfer - Catalysis Topic Area (multiple)	
Small Business Innovation Research/Small Business Technology Transfer - Advanced Solar Technology Topic Area (multiple)	
Department of Housing and Urban Development	
Neighborhood Stabilization Program 2 (Office of Community and Planning Development)	This was an American Recovery and Reinvestment Act of 2009 program and all grant funds were obligated as of February 11, 2010. Grantees must complete expenditure of these funds by February 2013.
Mark to Market Green Initiative (Office of Housing)	Statutory authority for this program ended at the end of fiscal year 2011.
Green Retrofit Program for Multifamily Housing (Office of Housing)	This was an American Recovery and Reinvestment Act of 2009 program and all funding was obligated as of September 30, 2010.
Capital Fund Recovery Act Competitive Grant Program (Office of Public and Indian Housing)	These grants were funded on a one-time basis in 2009 under theAmerican Recovery and Reinvestment Act of 2009.
Department of the Interior	
Desalination and Water Purification Research and Development Program (Bureau of Reclamation)	This initiative ended at the end of fiscal year 2011.
Geothermal Leasing and Operations on Federal Lands (Bureau of Land Management)	Authorization recently expired for the Geothermal Steam Act Implementation Fund, which was established by Section 234 of the Energy Policy Act of 2005 and has been used to expedite thedevelopment of geothermal steam as part of this initiative. Existing funding is expected to be depleted in 2012.
Renewable Energy Program: Multipurpose Marine Cadastre (Bureau of Ocean Energy Management, Regulation and Enforcement)	Authority for this initiative will expire at the end of fiscal year 2014.
External Renewable Energy Training (Indian Affairs)	These are one-time activities authorized under the American Recovery and Reinvestment Act of 2009.

Initiative (agency)	Expiration information
Housing Improvement Program: Recovery Act Renewable Energy Projects (Indian Affairs)	
Recovery Act Renewable Energy Efforts (Bureau of Land Management)	
Hydropower Optimization and Modernization (Bureau of Reclamation)	This initiative was funded through a one-time allocation, and will not be continued after funding runs out.
WaterSMART Grant Program (formerly known as Challenge Grants) (Bureau of Reclamation)	The authorizing legislation for this program includes an appropriations cap of $200 million, to remain available until expended.
Department of Transportation	
Urbanized Area Formula Program (Federal Transit Administration)	Authorization for these programs is contingent upon extension of the Safe, Accountable, Flexible, Efficient Transportation Equity Act of 2005: A Legacy for Users.
Transportation for Elderly Person and Persons with Disabilities (Federal Transit Administration)	
New Freedom Program (Federal Transit Administration)	
Major Capital Investments (New Starts & Small Starts) (Federal Transit . Administration)	
Job Access and Reverse Commute Program (Federal Transit Administration)	
Formula Grants for Other than Urbanized Areas (Federal Transit Administration)	
Fixed Guideway Modernization (Federal Transit Administration)	
Clean Fuels Grant Program (Federal Transit Administration)	
Bus and Bus Facilities (Federal Transit Administration)	
NextGen Environmental Research – Aircraft Technologies, Fuels and Metrics (Federal Aviation Administration)	The most recent extension of authority for the Federal Aviation Administration's activities extends through January 2012.[b]
NextGen – System Development – Environment and Energy (Federal Aviation Administration)	
Commercial Aviation Alternative Fuels Initiative (Federal Aviation Administration)	

Table 16. (Continued)

Initiative (agency)	Expiration information
Advanced Vehicle Technology Grants (Research and Innovative Technology Administration)	These initiatives were mandated under the Safe, Accountable, Flexible, Efficient Transportation Equity Act of 2005: A Legacy forUsers for fiscal years 2006 through 2009, and were extended through the Hiring Incentives to Restore Employment Act of 2010for fiscal year 2010 only.
Advanced Research Program - Biobased Transportation Research (Research and Innovative Technology Administration)	
Advanced Research Program - Alternative Fuels and Life Cycle Engineering (Research and Innovative Technology Administration)	This initiative was mandated under the Safe, Accountable, Flexible, Efficient Transportation Equity Act of 2005: A Legacy forUsers for fiscal years 2006 through 2009, but was not extended through the Hiring Incentives to Restore Employment Act of 2010.
Department of the Treasury	
Payments for Specific Energy Property in Lieu of Tax Credits (Office of Domestic Finance)[a]	Applications must be submitted before October 1, 2012, unless the Internal Revenue Code is amended.
Credit for Holding Clean Renewable Energy Bonds (Internal Revenue Service)	The allocation period for Clean Renewable Energy Bonds expired December 31, 2009.
Alcohol and Alcohol Fuel Mixtures Excise and Income Tax Credits (Internal Revenue Service)	These credits expire on December 31, 2011, unless the Internal Revenue Code is amended.[b]
Alternative Fuel Excise Tax Credit (Internal Revenue Service)	
Small Ethanol Producer Credit (Internal Revenue Service)	
Ethanol Excise and Income Tax Credits (Internal Revenue Service)	
Biodiesel Excise and Income Tax Credits (Internal Revenue Service)	
Small Agri-Biodiesel Producer Credit (Internal Revenue Service)	
Credit for the Production of Cellulosic Biofuel (Internal Revenue Service)	This credit expires on December 31, 2012, unless the Internal Revenue Code is amended.
Accelerated Depreciation Recovery Periods for Specific Energy Property (Internal Revenue Service)	Bonus depreciation was extended through December 31, 2012, by the Tax Relief, Unemployment Insurance Reauthorization, and Job Creation Act of 2010.

Initiative (agency)	Expiration information
Credit for Residential Energy Efficient Property (Internal Revenue Service)	Property must be placed in service by December 31, 2016.
Energy Production Credit (Internal Revenue Service)	Projects must be placed in service by either December 31, 2012, or December 31, 2013, to be eligible.
Credit for Clean Fuel Burning Vehicles (Internal Revenue Service)	Credits for clean-fuel burning hybrid, advanced lean-burn technology, and alternative fuel vehicles expired December 31, 2010; credits for plug-in electric vehicles expire December 31, 2011, unless the Internal Revenue Code is amended; credits for plug-in electric-drive vehicles expire December 31, 2014, unless the Internal Revenue Code is amended; and credits for fuel cell vehicles expire December 31, 2014, unless the Internal Revenue Code is amended.[b]
Credit for Clean-Fuel Burning Vehicle Refueling Property (Internal Revenue Service)	Unless the Internal Revenue Code is amended, the credit for qualified refueling property expires December 31, 2011, except for hydrogen refueling property, which expires December 31, 2014. The higher credit rate and credit cap expired December 31, 2010.[b]
Qualifying Advanced Energy Project Credit (Internal Revenue Service)	The application period for the first round of credit certification ended on December 31, 2010. All available amounts were allocated in 2010. Property certified in the first round must be placed in service within 3 years of certification (generally in 2014). If some credits which were allocated are forfeited, it may be possible to reallocate them through 2013.
Energy Investment Credit (Internal Revenue Service)	Wind projects must be placed in service by December 31, 2012, in order to claim the credit; other facilities specified under InternalRevenue Code Section 45 must be placed in service by December 31, 2013; solar electricity, heat, or light, geothermal electricity or heat, small wind, and other property specified under Internal Revenue Code Section 48 must be placed in service by December 31, 2016.
National Aeronautics and Space Administration	

Table 16. (Continued)

Initiative (agency)	Expiration information
National Aeronautics and Space Administration Third Party-Financed Renewable Power Generation	The on-site power generation initiative at the Kennedy Space Center was a pilot project during the time period when the National Aeronautics and Space Administration received temporary enhanced use lease authority, including in-kind consideration. The agency's authority to enter into enhanced use leases expires on December 26, 2017; for this initiative to continue, the National Aeronautics and Space Act of 1958 would need to be modified to extend or remove this sunset provision.
Small Business Administration	
Energy Grants for Green Technology	Authority to make grants under these programs ends in 2013.
Energy Efficiency Grants for Education and Technology	

Source: GAO analysis of agencies' data.

[a] The Department of the Treasury offers payments for specific energy property in lieu of either the Energy Production or Energy Investment tax credits.

b Agencies reported information on expiring initiatives as of fall of 2011 and we did not update this information to reflect whether or not any of the initiatives expected to expire prior to the issuance of our final report—such as certain tax expenditures that supported ethanol or other biofuels—actually did expire.

End Notes

[1] We defined renewable energy as energy derived from any of the following sources: bioenergy, such as liquid biofuel and solid biomass fuel; geothermal; hydropower; solar; wind; ocean energy, including wave, tidal, current, and ocean thermal energy; and waste conversion, including anaerobic digestion, landfill gas, and municipal solid waste.

[2] DOE, Energy Information Administration, *Renewable Energy Consumption and Electricity Preliminary Statistics 2010* (Washington, D.C.: 2011).

[3] Throughout this report we refer to federal departments and independent agencies, as well as some components of these departments and agencies, collectively as "agencies."

[4] Since DOE was established in 1977, one of its missions has been to promote the nation's energy security through research, development, and demonstration of advanced technologies for meeting future energy demands and diversifying the nation's energy portfolio. As part of this mission, EERE conducts research, development, and demonstration activities in partnership with industry to advance a diverse supply of what it refers to as "clean energy" technologies, including renewable energy.

[5] Office of Management and Budget guidance provided that agencies that received Recovery Act funding were responsible for developing program-specific plans to outline how they would implement the provisions of the act, including timeframes for obligating the funds provided. In September 2011, the Office of Management and Budget directed federal agencies to accelerate spending of remaining Recovery Act funds in all discretionary grant programs, and directed them to reclaim funds that have not been spent by September 30, 2013, to the extent permitted by law.

[6] A tax expenditure is a revenue loss attributable to a provision of the federal tax laws that (1) allows a special exclusion, exemption, or deduction from gross income or (2) provides a special credit, preferential tax rate, or deferral of tax liability. Tax expenditures are subsidies provided through the tax system. Rather than transferring funds from the government to the private sector, the U.S. government forgoes some of the tax revenues that it would have collected and the taxpayers that take advantage of the provisions of the tax code pay lower taxes than they would otherwise have had to pay. See also GAO, *Government Performance and Accountability: Tax Expenditures Represent a Substantial Federal Commitment and Need to Be Reexamined*, GAO-05-690 (Washington, D.C.: Sept. 23, 2005), and *Understanding the Tax Reform Debate: Background, Criteria, & Questions*, GAO-05-1009SP (Washington, D.C.: September 2005).

[7] While sufficiently reliable as a gauge of general magnitude, the sum of individual tax expenditure estimates does not take into account interactions between individual provisions.

[8] We have reported on agencies' responses to these directives on several occasions. See GAO, *Federal Energy Management: Agencies Are Taking Steps to Meet High-Performance Federal Building Requirements, but Face Challenges*, GAO-10-22 (Washington, D.C.: Oct. 30, 2009); *Status of GSA's Implementation of Selected Green Building Provisions of the Energy Independence and Security Act of 2007*, GAO-09-111R (Washington, D.C.: Oct. 31, 2008); *Federal Energy Management: Agencies Are Acquiring Alternative Fuel Vehicles but Face Challenges in Meeting Other Fleet Objectives*, GAO-09-75R (Washington, D.C.: Oct. 22, 2008); and *Federal Energy Management: Addressing Challenges through Better Plans and Clarifying the Greenhouse Gas Emission Measure Will Help Meet Long-term Goals for Buildings*, GAO-08-977 (Washington, D.C.: Sept. 30, 2008).

[9] In particular, GAO-12-259SP provides information on each initiative, including what it is, how it works, and how it relates to renewable energy.

[10] Renewable energy activities may be part of broader initiatives which are not primarily focused on renewable energy. In these instances, renewable energy projects can be one of many eligible types of activities that receive support under an initiative. For example, initiatives aimed at promoting sustainability, energy efficiency, fuel economy, and other "green" measures may include a renewable energy component.

[11] Where we identified individual renewable energy-related projects, such as a specific facility energy effort or grant award, we did not consider the projects to be initiatives themselves; rather, we identified the broader program or area of effort to which they belonged as an initiative. Therefore, the data that we developed do not necessarily represent initiatives that are of similar size in terms of agency financial commitments, number of projects involved, or other quantitative measures. Also, our data do not always match agencies' reported information on these activities, such as information contained in agency budget or strategic planning documents. Similarly, the list of tax expenditures included in this report was developed based on information provided by IRS officials to represent provisions of the Internal Revenue Code that provide support for renewable energy. The list was developed specifically to represent the criteria for collecting data on distinct renewable energy

initiatives for this report. As a result, the count does not necessarily track Treasury's published tax expenditure list which, for example, may encompass multiple provisions in a tax expenditure listed.

[12] The 24 federal agencies subject to the CFO Act accounted for roughly 98 percent of federal outlays in 2009.

[13] FERC is an independent agency that, while not a CFO Act agency, is listed under DOE in the federal budget.

[14] We included technology demonstration activities as part of research and development rather than as part of commercialization and deployment. In addition, we included equipment, such as tactical military equipment, that is procured and deployed by agencies and that incorporates renewable energy technologies as part of the fleets and facilities federal role.

[15] Liquid biofuels, such as ethanol or biodiesel, are fuels or components of fuels produced from feedstocks, including corn, sugar cane, algae, and a range of plant oils and animal fats, among others. Liquid biofuels are primarily used for transportation. We refer to biofuels, such as ethanol and biodiesel, as renewable energy sources in this report, although they are commonly referred to as alternative fuels. Solid biomass fuels are fuels derived from organic matter, such as wood and forest residues that can be burned to produce heat energy. Other types of bioenergy include biogas, such as methane. However, for purposes of our analysis we categorized gases produced from landfill gas capture or anaerobic digestion of agricultural waste as waste conversion rather than bioenergy.

[16] GAO reports on federal agencies' methods of financing energy projects include GAO, *Energy Savings: Performance Contracts Offer Benefits, but Vigilance Is Needed to Protect Government Interests*, GAO-05-340 (Washington, D.C.: June 22, 2005), and *Capital Financing: Partnerships and Energy Savings Performance Contracts Raise Budgeting and Monitoring Concerns,* GAO-05-55 (Washington, D.C.: Dec. 16, 2004).

[17] DOE, Energy Information Administration, *Direct Federal Financial Interventions and Subsidies in Energy in Fiscal Year 2010* (Washington, D.C.: 2011).

[18] Treasury's Payments for Specific Energy Property in Lieu of Tax Credits, established under Section 1603 of the Recovery Act, provided cash grants to eligible applicants who place specified energy property in service for use in a trade or business. Congressional Research Service, *Energy Tax Incentives: Measuring Value Across Different Types of Energy Resources* (Washington, D.C.: 2011).

[19] Environmental Law Institute, *Estimating U.S. Government Subsidies to Energy Sources: 2002-2008* (Washington, D.C.: 2009).

[20] The federal agencies within the scope of this report are organized along a range of different structures that include formal subagencies, bureaus, services, divisions, and offices, among other types of structures. For ease of reporting, we generally refer to all of these organizational elements as subagencies.

[21] One Office of the Chief Economist initiative is implemented jointly with USDA's National Institute of Food and Agriculture, which is part of the mission area of the Under Secretary for Research, Education, and Economics; therefore, the number of initiatives listed for USDA sums to 106.

[22] We generally grouped DOE's Power Marketing Administrations together for purposes of our analysis; however, they include the Bonneville Power Administration, the Southeastern Power Administration, the Southwestern Power Administration, and the Western Area Power Administration.

[23] One initiative was jointly implemented by subagencies under three Assistant Secretaries; therefore the number of initiatives listed for Interior sums to 84.

[24] We did not analyze the impact of initiatives on the renewable energy sector for this report.

[25] Recent estimates have indicated that the estimated forgone revenues associated with tax expenditures administered by IRS represented more than $8 billion in fiscal year 2010 in federal financial support for renewable energy. See, for example, DOE, Energy Information Administration, *Direct Federal Financial Interventions and Subsidies in Energy in Fiscal Year 2010* (Washington, D.C.: 2011), and Congressional Research Service, *Energy Tax Incentives: Measuring Value Across Different Types of Energy Resources* (Washington, D.C.: 2011).

[26] Under the act, the Renewable Fuel Standard applies to transportation fuel sold or introduced into commerce in the 48 contiguous states. However, the Administrator of the EPA is authorized, upon a petition from Alaska or Hawaii, to allow the Renewable Fuel Standard to apply in that state. On June 22, 2007, Hawaii petitioned EPA to opt into the Renewable Fuel Standard, and the Administrator approved that request. For the purposes of this report, statements that the Renewable Fuel Standard applies to U.S. transportation fuel refer to the 48 contiguous states and Hawaii. For more information on biofuels and the Renewable Fuel Standard, see GAO, *Biofuels: Challenges to the Transportation, Sale, and Use of Intermediate Ethanol Blends*, GAO-11-513 (Washington, D.C.: June 3, 2011), and *Biofuels: Potential Effects and Challenges of Required Increases in Production and Use*, GAO-09-446 (Washington, D.C.: Aug. 25, 2009).

[27] See appendix II for more information on the renewable energy sources supported by agencies' initiatives.

[28] The number of initiatives agencies implemented that support each energy source does not provide the only measure of agencies' support for these sources. For example, DOE is not included among the agencies with the largest number of initiatives supporting bioenergy, solar, or wind energy. Rather, DOE's initiatives supported a wide range of renewable energy sources, and the number of its initiatives that supported bioenergy, solar, or wind energy did not exceed those of other agencies.

[29] Solar photovoltaic technologies convert energy from sunlight directly into electricity, using arrays of solar panels. Concentrating solar power technologies use mirrors to focus the sun's light energy and convert it into heat to create electrical power. In solar hot water technology, a collector absorbs and transfers heat from the sun to water, which is stored in a tank until needed.

[30] Of the 11 agencies that participate in the Small Business Innovation Research program, 2 (the departments of Education and Health and Human Services) were not included within the scope of our review. In addition, not all of the agencies included in our review funded renewable energy research as part of their implementation of the program. For more information on the Small Business Innovation Research program, see GAO, *Small Business Innovation Research: SBA Should Work with Agencies to Improve the Data Available for Program Evaluation*, GAO-11-698 (Washington, D.C.: Aug. 15, 2011).

[31] Ground-source heat pumps, also known as geothermal heat pumps, are devices that take advantage of the relatively constant temperature immediately beneath the earth's surface to provide heat in the winter and air conditioning in the summer. During the winter, a heat pump transfers the heat of the ground to a fluid-filled set of coils and then pumps this fluid into the building. During the summer, heat pumps work in reverse, extracting heat from the building and transferring it to the ground.

[32] These tax expenditures include credits for taxpayers holding Clean Renewable Energy Bonds, New Clean Renewable Energy Bonds, and Qualified Energy Conservation Bonds. While there are differences between these credits in the amount of the credit available to taxpayers

and how authority to issue the bonds was allocated, among other differences, generally for these bonds, taxpayers may receive an income tax credit in lieu of interest payments from the issuers of the bonds. Issuing these bonds helps tax-exempt entities finance projects that produce electricity from renewable energy sources because, since bondholders receive credits from the federal government, bond issuers can borrow while paying little or no interest on their debt. For the New Clean Renewable Energy Bonds and Qualified Energy Conservation Bonds, bond issuers also have the option to receive a direct payment equivalent to and in lieu of the amount of the tax credit that would otherwise go to the bondholder. In cases where bond issuers elect to receive a direct payment, this option helps tax-exempt entities finance projects that produce electricity from renewable energy sources because it provides an incentive for investors to purchase the bonds, since the investors' returns would not depend upon having sufficient taxable income to utilize a tax credit.

[33] See appendix III for additional information on the extent to which the Recovery Act established, expanded, or modified initiatives.

[34] See appendix IV for additional information on expiring initiatives.

[35] Research and development includes research efforts that further the knowledge of or the ability to create, develop, or improve renewable energy technologies, as well as, for purposes of our analysis, technology demonstration. Fleets and facilities include efforts to incorporate renewable energy into an agency's property, buildings, vehicles, and other equipment. Commercialization and deployment includes activities that provide incentives for the implementation or promote the competitiveness of renewable energy technologies in the commercial market place. Regulation, permitting, and compliance include activities that ensure compliance- with federal laws and regulations regarding renewable energy production, use, technologies, or facilities.

[36] See appendix II for more detail on agencies with initiatives we placed in the "other" category. Our inclusion of initiatives in the "other" category does not imply that these initiatives were any less important than those in the four key federal roles included in our analysis. Initiatives in the "other" category may provide important support for renewable energy in a variety of ways, such as helping to promote the use of renewable energy technologies globally. We included initiatives in the "other" category only to distinguish between those initiatives that we grouped within the four key federal roles we analyzed and those that we did not include within one of the four federal roles.

[37] For more information on ARPA-E's research activities see GAO, *Department of Energy: Advanced Research Projects Agency-Energy Could Benefit from Information on Applicants' Prior Funding*, GAO-12-112 (Washington, D.C.: Jan. 13, 2012).

[38] Regional systems of bioenergy production include, among other elements, (1) deployment of regionally-appropriate dedicated energy crops; (2) implementation of regional feedstock production and logistics networks; (3) development of bioenergy conversion technologies that can accept a range of feedstocks; (4) regional marketing and distribution systems; and (5) regional sustainability analyses, data collection and management, and decision-making tools.

[39] With 10 fleets and facilities initiatives, the percentage of DOE's initiatives in this federal role, relative to DOE's other initiatives, was 11 percent. As a result, we did not highlight a fleet and facilities federal role for DOE in table 5, which represents agencies where 25 percent or more of their initiatives were in a particular federal role. However, we included DOE here as having a key role in fleet and facilities activities governmentwide because of its efforts under FEMP, through which DOE assists all agencies to include renewable energy in their fleets and facilities.

[40] This tax expenditure is known as the Volumetric Ethanol Excise Tax Credit. Other tax expenditures administered by IRS provide similar excise tax credits for renewable alternative fuels sold or blended, including for alternative fuels other than ethanol, methanol, and biodiesel (50 cents per gallon), alcohol and alcohol fuel mixtures (60 cents per gallon), and biodiesel ($1.00 per gallon). IRS also administers other tax expenditures that support the production of renewable alternative fuels by providing an additional 10 cent per gallon income tax credit on the first 15 million gallons of ethanol produced each year by both small ethanol and certain small biodiesel producers, and a $1.01 per gallon income tax credit for the production of cellulosic biofuels.

[41] FERC has an important role in regulating the sale and transmission of energy in general; however, FERC indicated that in large measure this role did not include a specific focus on renewable energy. FERC does have responsibility for licensing nonfederal hydropower facilities and has taken steps to ensure that renewable energy has equal access in the energy market. Since FERC did not identify itself as having a significant regulatory role with regard to renewable energy specifically, we did not identify it as a key agency for this federal role.

[42] On October 1, 2011, BOEMRE reorganized into two independent entities: the Bureau of Ocean Energy Management and the Bureau of Safety and Environmental Enforcement. The Bureau of Ocean Energy Management is responsible for managing development of the nation's off-shore resources in an environmentally and economically responsible way, and its activities include oversight of leasing, environmental studies, and economic analysis. The Bureau of Safety and Environmental Enforcement is responsible for enforcing safety and environmental regulations. However, given the timeframe of our work, we refer to BOEMRE in this report.

End Notes for Appendix I

[1] We refer to federal departments and independent agencies, as well as some subcomponents of these departments and agencies collectively as "agencies."

[2] Pub. L. No. 101-576, 104 Stat. 2838 (1990). Some federal agencies not subject to the CFO Act may also implement renewable energy-related activities. For example, one function of the Tennessee Valley Authority is to market hydropower. Alternatively, the Export-Import Bank of the United States provides loans and other types of financing arrangements to help international buyers purchase U.S. goods and services, such as renewable energy equipment. We generally excluded the efforts of these and other agencies not subject to the CFO Act from the scope of our review.

[3] The Federal Energy Regulatory Commission is an independent agency that, while not a CFO Act agency, is listed under DOE in the federal budget.

[4] The CFO Act included the Federal Emergency Management Agency as one the 24 agencies subject to its provisions. However, in fiscal year 2005, the Department of Homeland Security became subject to the provisions of the CFO Act and the Federal Emergency Management Agency became part of the department.

[5] Given the large number of Department of State (State) embassies and consulates overseas, the initiatives for which we obtained information from State represent only a portion of State's renewable energy efforts. However, State indicated that they are representative of the types of renewable energy initiatives that are underway at its missions around the world.

[6] Given the large number of U.S. Agency for International Development field offices overseas, the initiatives for which we obtained information from the agency represent only a portion of its renewable energy efforts. However, the U.S. Agency for International Development indicated that they are representative of the types of renewable energy initiatives that are underway at its missions around the world.

[7] We did not include hydrogen as a renewable energy source, although we did consider efforts to produce hydrogen using renewable energy to be a renewable energy-related activity. Also, we did not consider technologies, such as fuel cells, to be a renewable energy source, although we did consider the use of renewable fuels as the power source for fuel cells to be a renewable energy-related activity.

[8] Renewable energy activities may be part of broader initiatives the primary focus of which is not renewable energy. In these instances, renewable energy projects can be one of many eligible types of activities that receive support under an initiative. For example, initiatives aimed at promoting sustainability, energy efficiency, fuel economy, and other "green" measures may include a renewable energy component.

[9] In addition, our data do not always match agencies' reported information on these activities, such as information contained in agency budget or strategic planning documents. Similarly, the list of tax expenditures included in this report was developed based on information provided by IRS officials to represent provisions of the Internal Revenue Code that provide support for renewable energy. The list was developed specifically to represent the criteria for collecting data on distinct renewable energy initiatives for this report. As a result, the count does not necessarily track Treasury's published tax expenditure list which, for example, may encompass multiple provisions in a tax expenditure listed.

[10] We examined fiscal year 2011 budget documents to help identify fiscal year 2010 activities because these documents provided information on how fiscal year 2010 funding was used.

[11] We generally classified the direct recipient of an initiative based on the initiative's function (i.e., who received the funding to conduct research, install a renewable energy project at an agency facility, or deploy a renewable energy technology; or, who was regulated by the initiative). For example, in many instances, federal agencies work through the General Services Administration or DOD's Defense Logistics Agency in implementing their fleets and facilities efforts. For purposes of our review, we did not consider these federal agencies to be recipients, but rather "pass-through" entities. We considered the recipients of agencies' fleets and facilities efforts to be the vehicle and equipment manufacturers, fuel suppliers, and contractors that the General Services Administration and Defense Logistics Agency use to serve their federal agency customers.

End Notes for Appendix II

[1] See appendix I for more information on our methodology and how we defined the key federal roles and "other."

[2] In commenting on a draft of this report, NSF indicated that its research activities are integrated with education in order to enhance the development of a world-class scientific and engineering workforce.

End Notes for Appendix III

[1] Some initiatives were affected by the Recovery Act in multiple ways.

[2] The Department of the Treasury offered payments for specific energy property in lieu of the Energy Production and Energy Investment tax credits.

In: Renewable Energy Initiatives Supported ... ISBN: 978-1-62081-645-5
Editors: R. Taylor and A. E. Martin

Chapter 2

RENEWABLE ENERGY R&D FUNDING HISTORY: A COMPARISON WITH FUNDING FOR NUCLEAR ENERGY, FOSSIL ENERGY, AND ENERGY EFFICIENCY R&D*

Fred Sissine

SUMMARY

Energy research and development (R&D) intended to advance technology played an important role in the successful outcome of World War II. In the post-war era, the federal government conducted R&D on fossil fuel and nuclear energy sources to support peacetime economic growth. The energy crises of the 1970s spurred the government to broaden the focus to include renewable energy and energy efficiency. Over the 33-year period from the Department of Energy's inception at the beginning of fiscal Year (FY) 1978 through FY2010, federal spending for renewable energy R&D amounted to about 16% of the energy R&D total, compared with 14% for energy efficiency, 26% for fossil, and 37% for nuclear. For the 63-year period from 1948 through 2010, nearly 12% went to renewables, compared with 9% for efficiency, 25% for fossil, and 50% for nuclear.

---

* This is an edited, reformatted and augmented version of a Congressional Research Service publication, CRS Report for Congress RS22858, from www.crs.gov, dated January 26, 2011.

Introduction

This report provides a cumulative history of Department of Energy (DOE) funding for renewable energy compared with funding for the other energy technologies—nuclear energy, fossil energy, and energy efficiency. Specifically, it provides a comparison that covers cumulative funding over the past 10 years (FY2001-FY2010), a second comparison that covers the 33-year period since DOE was established at the beginning of fiscal year 1978 (FY1978-FY2010), and a third comparison that covers a 63-year funding history (FY1948-FY2010).

Guide to Tables and Charts

Table 1 shows the cumulative funding totals in real terms for the past 10 years (first column), 33 years (second column), and 63 years (third column). Table 2 converts the data from Table 1 into relative shares of spending for each technology, expressed as a percentage of total spending for each period.

Figure 1 displays the data from the first column of Table 2 as a pie chart. That chart shows the relative shares of cumulative DOE spending for each technology over the 10 years from FY2001 through FY2010. Figure 2 provides a similar chart for the period from FY1978 through FY2010. Figure 3 shows a chart for FY1948 through FY2010.

Background

The availability of energy—especially gasoline and other liquid fuels—played a critical role in World War II. Another energy-related factor was the application of research and development (R&D) to the atomic bomb and other military technologies. During the post World War II era, the federal government began to apply R&D to the peacetime development of energy sources to support economic growth. At that time, the primary R&D focus was on fossil fuels and new forms of energy derived from nuclear fission and nuclear fusion.

From FY1948 through FY1977 the federal government provided an extensive amount of R&D support for fossil energy and nuclear power technologies.[1] Total spending on fossil energy technologies over that period amounted to about $15.7 billion, in constant FY2010 dollars. The federal government spent about $47.5 billion (in constant FY2010 dollars) during that period for nuclear fission and nuclear fusion energy R&D.[2]

The energy crises of the 1970s spurred the federal government to expand its R&D programs to include renewable (wind, solar, biomass, geothermal, hydro) energy and energy efficiency technologies. Modest efforts to support renewable energy and energy efficiency began during the early 1970s. From FY1973 through FY1977 the federal government spent about $1.6 billion (in constant FY2010 dollars) on renewable energy R&D, $143 million on energy efficiency R&D, and $173 million on electric systems R&D.[3]

The Department of Energy was established by law in 1977. All of the energy R&D programs— fossil, nuclear, renewable, and energy efficiency— were brought under its administration. DOE also undertook a small program in energy storage and electricity system R&D that supports the four main energy technology programs.[4] DOE's funding support for those technologies began in FY1978. Funding for all four of the main technologies skyrocketed initially, and then fell dramatically in the early 1980s.

Table 1. DOE Energy Technology Cumulative Funding Totals (billions of 2010 dollars)

	Period		
Technology	FY2001-FY2010 (10 years)	FY1978-FY2010 (33 years)	FY1948-FY2010 (63 years)
Renewable Energy	$ 6.42	$ 19.50	$ 21.06
Energy Efficiency	6.50	16.87	17.02
Fossil Energy	10.55	30.87	46.61
Nuclear Energy	8.85	43.63	91.12
Electric Systems	5.78	7.92	8.09
Total	$38.10	$118.79	$183.91

Sources: DOE Budget Authority History Table by Appropriation, May 2007; DOE Congressional Budget Requests (several years); DOE (Pacific Northwest Laboratory), An Analysis of Federal Incentives Used to Stimulate Energy Production, 1980. Deflator Source: The Budget for Fiscal Year 2011. Historical Tables. Table 10.1. Gross Domestic Product and Deflators Used in the Historical Tables, 1940-2015.

Table 2. DOE Energy Technology Share of Funding (percent; derived from Table 1)

Technology	Period		
	FY2001-FY2010 (10 years)	FY1978-FY2010 (33 years)	FY1948-FY2010 (63 years)
Renewable Energy	16.8%	16.4%	11.5%
Energy Efficiency	17.1%	14.2%	9.3%
Fossil Energy	27.7%	26.0%	25.3%
Nuclear Energy	23.2%	36.7%	49.5%
Electric Systems	15.2%	6.7%	4.4%
Total	100.0%	100.0%	100.0%

Sources: DOE Budget Authority History Table by Appropriation, May 2007; DOE Congressional Budget Requests (several years); DOE (Pacific Northwest Laboratory), An Analysis of Federal Incentives Used to Stimulate Energy Production, 1980; DOE Conservation and Renewable Energy Base Table. February 1990. Deflator Source: The Budget for Fiscal Year 2011. Historical Tables. Table 10.1. Gross Domestic Product and Deflators Used in the Historical Tables, 1940-2015.

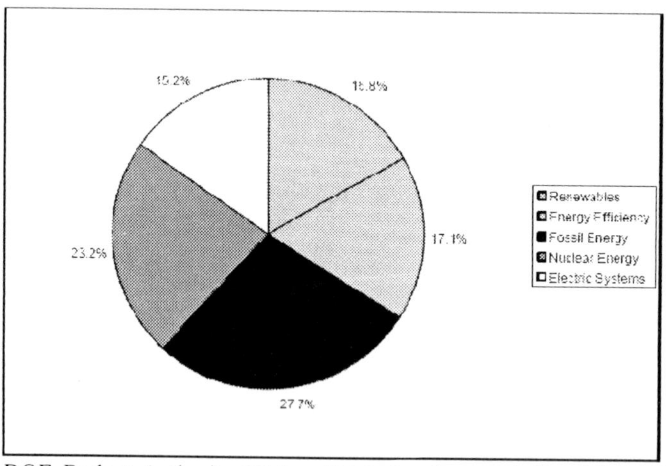

Sources: DOE Budget Authority History Table by Appropriation, May 2007; DOE Congressional Budget Requests (several years); Deflator Source: The Budget for Fiscal Year 2011. Historical Tables. Table 10.1. Gross Domestic Product and Deflators Used in the Historical Tables, 1940-2015.

Figure 1. DOE Energy Technology Share of Funding, FY2001-FY2010.

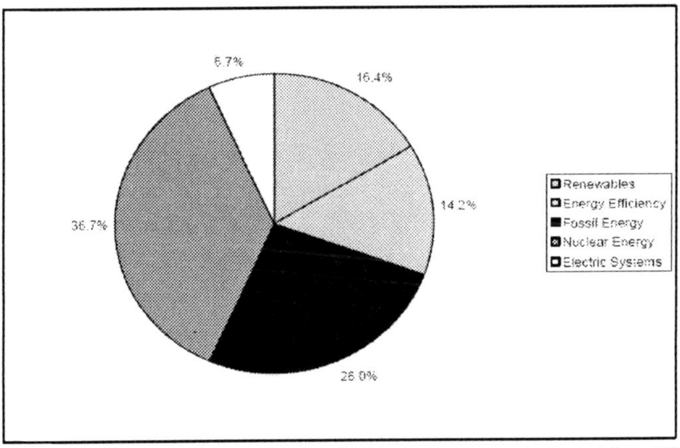

Sources: DOE Budget Authority History Table by Appropriation, May 2007; DOE Congressional Budget Requests (several years); Deflator Source: The Budget for Fiscal Year 2011. Historical Tables. Table 10.1. Gross Domestic Product and Deflators Used in the Historical Tables, 1940-2015.

Figure 2. DOE Energy Technology Share of Funding, FY1978-FY2010.

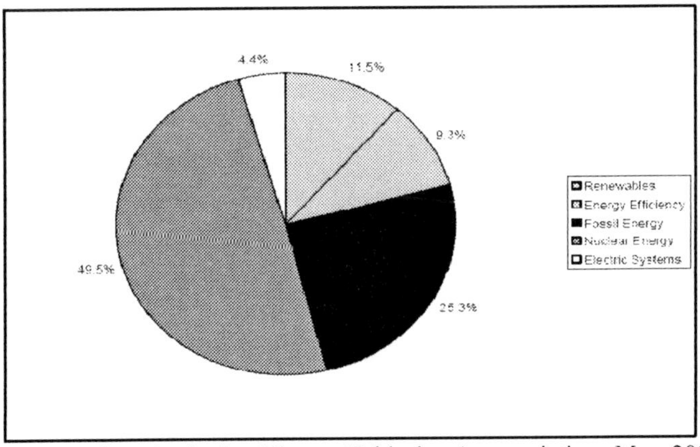

Sources: DOE Budget Authority History Table by Appropriation, May 2007; DOE Congressional Budget Requests (several years); DOE (Pacific Northwest Laboratory), An Analysis of Federal Incentives Used to
Stimulate Energy Production, 1980; DOE Conservation and Renewable Energy Base Table. Feb. 1990. Deflator Source: The Budget for Fiscal Year 2011. Historical Tables. Table 10.1.

Figure 3. DOE Energy Technology Share of Funding, FY1948-FY2010.

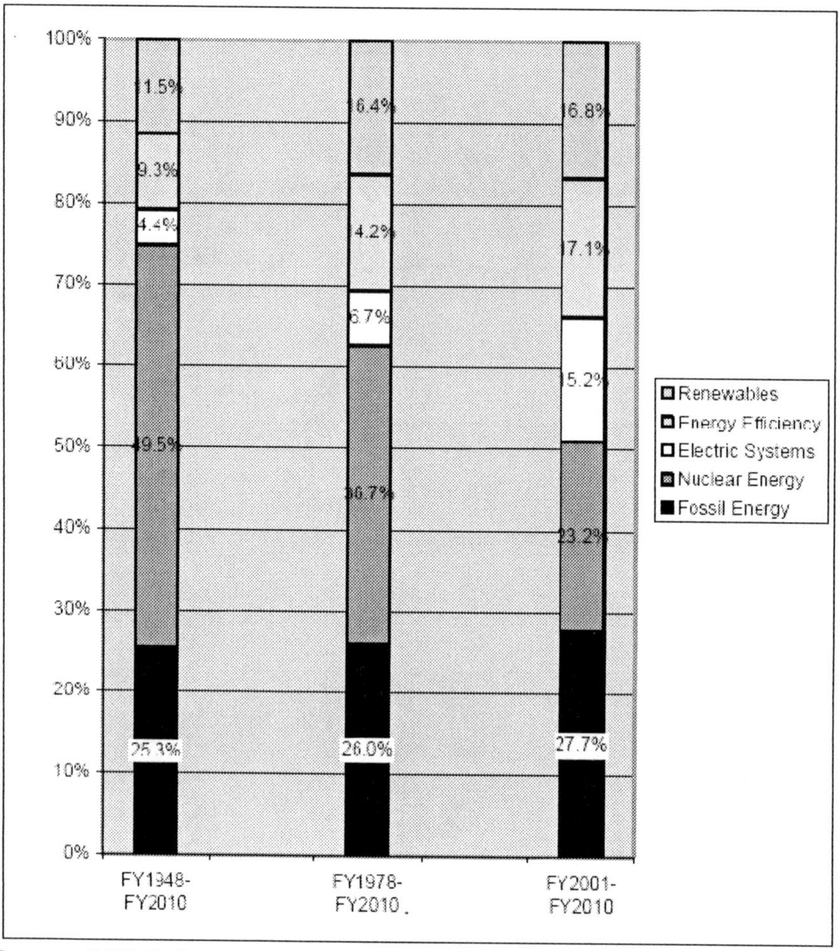

Sources: DOE Budget Authority History Table by Appropriation, May 2007; DOE Congressional Budget Requests (several years); DOE (Pacific Northwest Laboratory), An Analysis of Federal Incentives Used to Stimulate Energy Production, 1980; DOE Conservation and Renewable Energy Base Table. Feb. 1990. Deflator Source: The Budget for Fiscal Year 2011. Historical Tables. Table 10.1.

Note: Column to far left shows shares for the period FY1948-FY2010; middle column shows shares for period.from FY1978-FY2010; and far right column shows shares for period from FY2001-FY2010.

Figure 4. DOE Energy Technology Share of Funding, Comparison over Three Periods.

End Notes

[1] DOE. Pacific Northwest Laboratory. An Analysis of Federal Incentives Used to Stimulate Energy Production. 1980. The spending for fossil energy included coal, oil, and natural gas technologies.

[2] DOE (Pacific Northwest Laboratory), An Analysis of Federal Incentives Used to Stimulate Energy Production, 1980.

[3] DOE Conservation and Renewable Energy Base Table. February 1990.

[4] This program includes R&D on advanced batteries to store electricity and transmission equipment to transfer electricity with less heat loss (i.e. at higher levels of energy efficiency).

INDEX

C

D

E

F

G

S

T

U

V

W